Oxford Chemistry Series

General Editors
P. W. ATKINS J. S. E. HOLKER A. K. HOLLIDAY

Oxford Chemistry Series

1972
1. K. A. McLauchlan: *Magnetic resonance*
2. J. Robbins: *Ions in solution (2): an introduction to electrochemistry*
3. R. J. Puddephatt: *The periodic table of the elements*
4. R. A. Jackson: *Mechanism: an introduction to the study of organic reactions*

1973
5. D. Whittaker: *Stereochemistry and mechanism*
6. G. Hughes: *Radiation chemistry*
7. G. Pass: *Ions in solution (3): inorganic properties*
8. E. B. Smith: *Basic chemical thermodynamics*
9. C. A. Coulson: *The shape and structure of molecules*
10. J. Wormald: *Diffraction methods*
11. J. Shorter: *Correlation analysis in organic chemistry: an introduction to linear free-energy relationships*
12. E. S. Stern (ed): *The chemist in industry (1): fine chemicals for polymers*
13. A. Earnshaw and T. J. Harrington: *The chemistry of the transition elements*

1974
14. W. J. Albery: *Electrode kinetics*
16. W. S. Fyfe: *Geochemistry*
17. E. S. Stern (ed): *The chemist in industry (2): human health and plant protecti*
18. G. C. Bond: *Heterogeneous catalysis: principles and applications*
19. R. P. H. Gasser and W. G. Richards: *Entropy and energy levels*
20. D. J. Spedding: *Air pollution*
21. P. W. Atkins: *Quanta: a handbook of concepts*
22. M. J. Pilling: *Reaction kinetics*
23. J. N. Bradley: *Fast reactions*
24. M. H. Freemantle: *The chemist in industry (3): management and economics*

1976
25. C. F. Bell: *Principles and applications of metal chelation*
26. D. A. Phipps: *Metals and metabolism*

D. A. PHIPPS
Department of Chemistry, Liverpool Polytechnic

Metals and metabolism

Clarendon Press · Oxford

Oxford University Press, Walton Street, Oxford OX2 6DP

OXFORD LONDON GLASGOW
NEW YORK TORONTO MELBOURNE WELLINGTON
KUALA LUMPUR SINGAPORE JAKARTA HONG KONG TOKYO
DELHI BOMBAY CALCUTTA MADRAS KARACHI
IBADAN NAIROBI DAR ES SALAAM CAPE TOWN

Paperback ISBN 0 19 855413 3
Casebound ISBN 0 19 855452 4

First published 1976
Reprinted (with corrections) 1978

Set by Hope Services Ltd., Wantage
Printed in Great Britain
by J. W. Arrowsmith Ltd., Bristol

Editor's Foreword

Although the importance of metal complexes in both plants and animals has been recognized for many years, the chemistry of living systems has tended to be related to the organic rather than to the inorganic area of chemistry. The great advances in the study not only of metal complexes but more generally of metal ions in solution has altered this situation, and the importance of inorganic chemistry in relation to metabolic processes is now recognized.

This book sets out to relate, as far as is possible, the biological role of each metal to its position in the periodic table and to its appropriate 'inorganic' properties.

In some cases, this relationship is fairly well understood; in others it is, in the present state of our knowledge, decidedly tenuous. It is hoped that these contrasts will both inform and encourage the reader in an area where there is enormous scope for future development.

A. K. H.

Preface

TO many people, the progress of science appears to be a process of fission rather than of fusion, in which the established disciplines fragment into a bewildering array of 'specialisms', arising from the limits that time imposes on each of us. Such an apparent narrowing of interest and outlook can be seen in the schisms within the broad field of biochemistry, which has recently produced yet another new area of specialization, bio-inorganic chemistry.

A close examination of this new domain, with its etymologically horrendous name, serves as a timely reminder that science may make the most successful match from the least likely partners. For many years biochemistry has been seen principally as an outgrowth of its parent, organic chemistry, and the major part of the literature has dealt with the biological implications of the properties and reactions of compounds of carbon, ranging from very small molecules of only a few atoms to the largest natural macromolecules. More recently, however, interest has been rapidly increasing in the biological roles of other elements, particularly metals; this upsurge has been preceded and aided by the rapid advances in inorganic chemistry over the last twenty years. The result is that bio-inorganic chemistry is now an established and recognized field of study. Yet the process should not be seen as a narrowing of our outlook for, far from having a 'blinkering' effect, the growth of this new area has provided both a new viewpoint on existing knowledge and the tantalizing prospect of entirely new vistas.

Unlike the older, better-established disciplines the literature of bio-inorganic chemistry is not conveniently collected in a few major journals but is scattered about and is to be found anywhere and almost everywhere: not only in the journals of agriculture and of botany and, of course, in texts on chemistry, but also in the works on dentistry, geochemistry, medicine, nutrition, physiology, and zoology. It is a salutory lesson for the newcomer to this diversity to see the difference in emphasis and approach which specialist viewpoints bring to the same problem. This is really one of the many attractions of this new subject area; it enables us to go wandering off in fields which we might once have considered unconnected with either metals or metabolism. This freedom to browse through the fruits of other peoples' efforts is not without its price, since we must first take the trouble to learn at least the basic elements of the professional languages of the many workers from whose efforts we are to borrow so freely.

This short book represents a chemist's view of biological events, and in producing it I have had to rely heavily on the good-natured cooperation of my colleagues, who have endeavoured to correct my many misconceptions as I have trampled through fields other than my own. In order that the new reader shall not be overwhelmed I have endeavoured to explain technical terms as they have arisen in the text, and I have also suggested some other sources to which the interested reader can go for further clarification.

Inevitably in so brief a survey of such a wide-ranging subject there are many omissions which some will consider of major importance and inclusions which others will regard as trivia. More important, the choice of subject matter is largely composed of areas of the subject for which we can put forward some chemical explanation for biological actions, and the reader should be aware that, despite what we might like to think at the present, such cases are the exception rather than the rule. And it is to exceptions that our attention should in future be directed.

Liverpool D.A.P.

Contents

1. Introduction

Metals, metallic elements, and metal ions

AT the present time over 100 elements have been isolated, of which about 80 are usually described as metals.† The classification as a metal is based largely on a consideration of physical properties, and implies that the element has certain characteristics such as lustre, high density, large electrical and thermal conductivity, high melting point or considerable mechanical strength. All these characteristics are the recognizable attributes of familiar metals such as iron, but as is often the case with supposedly precise scientific definitions, this division of the elements into metals and non-metals is somewhat arbitrary, since the characteristics of each group are not mutually exclusive. Some elements, which by common consent belong quite definitely to one class, may well have properties reminiscent of the other. The low melting points of the alkalis, which are otherwise typical metals, and the high electrical conductivity of carbon (as graphite), a non-metal, are obvious examples of where difficulty in classification may arise. Indeed, in several cases it has proved impossible to place an element satisfactorily in either class, and elements such as boron and silicon have to be classified separately as metalloid or metal-like.

One disadvantage of our classification scheme is that it focuses attention on physical characteristics whilst ignoring chemical properties; though with such a large number of metallic elements, each with its own rich and varied chemistry, it would obviously be difficult to arrive at a comprehensive yet useful chemical classification. Nevertheless, such perspective is needed, since in biochemical or geochemical studies it is seldom that either physical properties or even the elemental state itself are considered, because with few exceptions the metallic elements do not occur in the native state but are found in salts or complexes of the positively charged metal ion.

Perhaps, then, a limited but useful definition of a metal is that it is an element which under biologically significant conditions may react by losing one or more electrons to form a cation, this being the functionally significant species. Of course, this is not meant to imply that metals may not react in a variety of other ways which do not fit this definition, but simply that these other reactions are normally limited to distinctly abiotic conditions.

† At least 103 elements have been definitely identified, and this number are included in most modern periodic tables.

Typical elements					H	He			
	1s				1	2			
		Li	Be	B	C	N	O	F	Ne
He core +	2s	1	2	2	2	2	2	2	2
	2p			1	2	3	4	5	6
		Na	Mg	Al	Si	P	S	Cl	Ar
Ne core +	3s	1	2	2	2	2	2	2	2
	3p			1	2	3	4	5	6

		K	Ca	Sc	Ti	V	Cr	Mn	Fe	Co	Ni	Cu	Zn	Ga	Ge	As	Se	Br	Kr
Ar core +	3d			1	2	3	5	5	6	7	8	10	10	10	10	10	10	10	10
	4s	1	2	2	2	2	1	2	2	2	2	1	2	2	2	2	2	2	2
	4p													1	2	3	4	5	6
		Rb	Sr	Y	Zr	Nb	Mo	Tc	Ru	Rh	Pd	Ag	Cd	In	Sn	Sb	Te	I	Xe
Kr core +	4d			1	2	4	5	5	7	8	10	10	10	10	10	10	10	10	10
	5s	1	2	2	2	1	1	2	1	1	0	1	2	2	2	2	2	2	2
	5p													1	2	3	4	5	6
		Cs	Ba	Lanthanides	Hf	Ta	W	Re	Os	Ir	Pt	Au	Hg	Tl	Pb	Bi	Po	At	Rn
Xe core +	4f				14	14	14	14	14	14	14	14	14	14	14	14	14	14	14
	5d				2	3	4	5	6	7	9	10	10	10	10	10	10	10	10
	6s	1	2		2	2	2	2	2	2	1	1	2	2	2	2	2	2	2
	6p													1	2	3	4	5	6
		Fr	Ra	Actinides		*A elements*									*B elements*				
Rn core +	6d																		
	7s	1	2																
Group		I	II	III	IV	V	VI	VII A		VIII		I	II	III	IV	V	VI	VII B	O

Lanthanides:	La	Ce	Pr	Nd	Pm	Sm	Eu	Gd	Tb	Dy	Ho	Er	Tm	Yb	Lu
Actinides:	Ac	Th	Pa	U	Np	Pu	Am	Cm	Bk	Cf	Es	Fm	Md	No	Lw

FIG. 1. The periodic table of the elements.

Periodicity

Happily, whilst each metallic element is chemically distinct from the others, the differences are not random, and there is in fact considerable regularity or periodicity in the variation of their properties. Many relationships (both similarities and differences) are emphasized when the elements are arranged in the form of a periodic table such as is shown in Fig. 1. Such tables, based in principle on the original suggested by Mendeleev, are of great assistance in rationalizing and clarifying the enormous amount of experimental data now available. Some very ingenious tables have been devised using properties both of the elements themselves and also of their compounds. One table, for example, was based solely on the colours of various ions, whilst another was devised by considering solubility properties. Now, of course, it is appreciated that chemical periodicity is simply a reflection of electronic configuration, and it is on the basis of their electronic structure that the chemical and biochemical properties of the metals will be discussed.

The formation of metal ions

The energy change involved in the formation of a metal cation by removal of electrons from the neutral atom depends critically on the immediate environment of the atom. Taking the simplest case first – the isolated atom in the gas phase – the energy required (the ionization energy) can easily be determined experimentally. Not unexpectedly it varies in a periodic fashion.

A brief review of the ionization energies of the elements reveals that high values, implying electronic stability, are particularly associated with atoms and ions having filled sets of orbitals. This is the basis of a simple and successful valency theory, which explains most of the reactions of the s- and p-block metals solely on the basis of their propensity to form cations with a closed-shell configuration. These ions are particularly stable and generally do not undergo any further changes in electronic configuration, so that the ionic charge is largely independent of the local environment. In contrast, the d- and f-block transition metals readily form several ions, each of different electronic configuration and mostly not conforming to the simple closed-shell rule. The stability of these ions depends critically on their immediate environment, as is recognized by the more sophisticated valency theories used to discuss their behaviour. A knowledge of the factors affecting the relative stability of each oxidation state is of fundamental importance in understanding their biochemical roles.

In contrast to reactions in the gas phase – for which considerable amounts of energy, perhaps of the order of several thousand kilojoules per mole, are required – the formation of a metal ion in aqueous solution is much more economical. Even where the formation of the ion is relatively unfavourable, the reduction in the overall energy change is still considerable. With copper, for example, the first ionization energy is more than 700 kJ mol^{-1} which is over 20 times the energy required to form Cu^+ in aqueous solution. In more favourable cases energy is actually released by the formation of the metal ion in water. A simplified explanation is provided in Fig.2. This shows that in solution the free energy of formation of the metal ion depends principally on the difference between two large factors, the ionization and the hydration energy; it is the balance between these which ultimately decides how many electrons will be lost from a particular atom. (Similar arguments may be advanced for the formation of ionic solids, in which the lattice energy is the counterpart of the hydration energy in solution.)

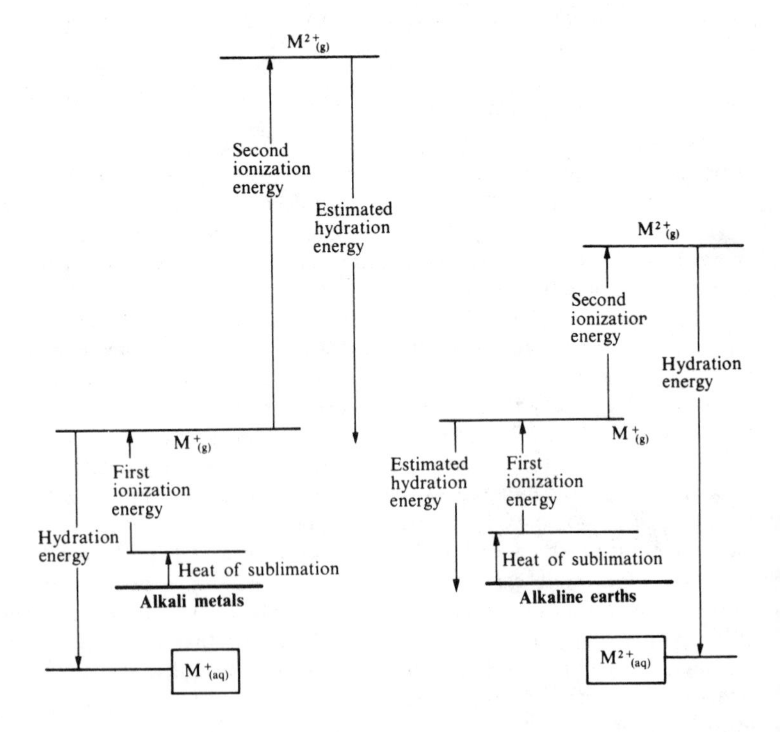

FIG. 2. The relative energy changes involved in the formation of aquated metal ions. In the case of the alkali metal the removal of more than one electron is energetically disallowed, whilst for the alkaline earth the formation of the dispositive ion is the most favoured process. Similar diagrams can be drawn for other metal ions.

Redox potentials

Since the biotic milieu is predominantly aqueous (though as will be seen later the local environment of a metal ion in a living system may be effectively non-aqueous) it is pertinent to continue by examining in slightly more detail the formation and properties of the metal ion in aqueous solution. The formation is considered not because there are many significant biochemical equilibria between the metal and its ions (though a few examples are known, and are of course important in geochemical cycling) but rather because of the information such equilibria provide about the relative stability of the ions.

The oxidation reaction (removal of electrons) of solid metal:

$$M_{(s)} \longrightarrow [M(OH_2)_x]^{n+} + ne^-$$

is just the process which occurs at the anode of an electrochemical cell. Unfortunately, however, it is impossible to make any measurements on such a reaction in isolation, since in practice it must always be accompanied by a corresponding reduction reaction, as, for example, in the Daniell cell.

$$Zn_{(s)} - 2e \longrightarrow Zn^{2+}_{(aq)} \quad \text{oxidation}$$

$$Cu^{2+}_{(aq)} + 2e \longrightarrow Cu_{(s)} \quad \text{reduction}$$

$$Zn_{(s)} + Cu^{2+}_{(aq)} \longrightarrow Zn^{2+}_{(aq)} + Cu_{(s)}, \quad E^{\ominus} = 1{\cdot}09\ V$$

where $E^{\ominus}$ is the e.m.f. at a defined standard state. Nevertheless it is still desirable to have some formal estimate of the contribution of each half of the reaction to the overall electromotive force of the cell. This may be done by combining each half-cell separately with a hydrogen electrode under carefully defined standard conditions and measuring the e.m.f. of each combination. The e.m.f. of the original cell may then be calculated by re-combining these two results, and since the contribution of the hydrogen cell is always removed arithmetically, it may conveniently (though arbitrarily) be set at zero.

$$H_2(g) \,\vdots\, H^+_{(aq)} \,|\, Zn^{2+}_{(aq)} \,\vdots\, Zn_{(s)}, \quad E^{\ominus} = -0{\cdot}76\ V$$

$$H_2(g) \,\vdots\, H^+_{(aq)} \,|\, Cu^{2+}_{(aq)} \,\vdots\, Cu_{(s)}, \quad E^{\ominus} = +0{\cdot}33\ V$$

whence $E^{\ominus} = +\,0{\cdot}33\ V-(-)0{\cdot}76\ V,$

$E^{\ominus} = +\,1{\cdot}09\ V.$

The e.m.fs of the two cells containing the hydrogen half-cell are then known as standard electrode potentials, or *redox* potentials.† In this fashion it is possible to set up a scale of potentials relative to the hydrogen electrode, which may subsequently be used to predict the course of reactions in solution, since the free energy change $\Delta G^{\ominus}$, is related to the cell e.m.f. E, and hence to the redox potentials themselves, by

$$\Delta G^{\ominus} = -nFE^{\ominus}$$

where n is the number of equivalents of electrons transferred during the course of the reaction and F is the Faraday.

Consequently, the information available from the redox potential may then be used to provide chemical information in a variety of situations. The redox potential for a single couple can first of all be used as guide to the relative stability of the element with respect to its cation. Thus if $E^{\ominus}$ for the couple $M^{n+}_{(aq)}/M_{(s)}$ is large and negative, as for the alkali metals, then the formation of the metal from its ion is most unfavourable, and conversely the formation of the cation is favoured. In other words, the metal is a powerful reducing agent, losing electrons easily, which is in accord with chemical experience. Clearly this argument may be reversed, and the more positive $E^{\ominus}$ becomes, the more powerful an oxidizing agent is the metal ion, more readily accepting electrons to return to the elemental state. However, a cautionary note is necessary; conclusions drawn in this fashion must be treated carefully, since these reactions cannot occur in isolation, and substances which act as oxidizing agents under one set of experimental conditions may well be reducing agents under others. For example, in the reaction below, the chromium functions first as a reducing agent and then as an oxidizing agent.

$$Cr^{III} \xrightarrow[OH^-]{H_2O_2} \begin{matrix} Cr^{VI} \\ + \\ H_2O \end{matrix} \xrightarrow[H^+]{H_2O_2} \begin{matrix} Cr^{III} \\ + \\ O_2 \end{matrix}$$

For a more detailed discussion of these reactions see p. 91.

Not surprisingly then confusion is common in the use of the terms oxidizing/reducing reagent and oxidation/reduction reaction. The reaction itself is termed oxidation or reduction *only* with respect to one of the reactants, and this should always be specified since in the course of a redox reaction the oxidizing agent is reduced whilst the reducing agent

† Considerable confusion exists because of the use of two opposite (European and American) sign conventions for treating electrochemical data. In this book redox potentials will be understood to conform to IUPAC convention and are written $M^{n+} + ne \longrightarrow M$, which are formally reduction potentials.

is itself oxidized. Thus the reaction between Cr^{III} and H_2O_2 can be styled either as the oxidation of Cr^{III} or the reduction of hydrogen peroxide.

However, whichever terminology is adopted, the overall course of the reaction may readily be predicted by considering the appropriate redox potentials. Moreover, redox reactions are by no means limited to metal-ion/metal couples and may equally apply to other biologically significant redox processes, for example,

$$\text{fumarate} + 2H^+ + 2e^- \rightarrow \text{succinate}, \quad E^\ominus = +0{\cdot}030 \text{ V},$$

$$\text{acetate} + 2H^+ + 2e^- \rightarrow \text{acetaldehyde}, \quad E^\ominus = -0{\cdot}600 \text{ V}.$$

From these values it is easy to deduce that fumarate is the more powerful oxidizing agent,† and acetaldehyde the more powerful reducing agent so that the reation becomes

$$\text{fumarate} + \text{acetaldehyde} \xrightarrow[\text{etc.}]{\text{enzymes; coenzymes}} \text{succinate} + \text{acetate}.$$

Clearly, a knowledge of redox potentials enables the thermodynamic probability of any reaction to be estimated, though of course no deduction can be made about the rate of such a reaction.

A second warning must now be given. Before redox potentials can be used to predict or explain the course of a reaction it is essential to ensure that the data being employed are applicable to the reaction conditions. The standard values $E^\ominus$ apply only when all reactants have unit activity, which is not necessarily nor even likely to be true *in vivo*.‡ As a consequence the actual redox potential E_{exp} will differ from the standard value, often quite markedly. For the alkali metals and alkaline earths even quite large changes in conditions such as concentration or temperature are unlikely to affect the stability of the metal ions, and it is still possible to ignore other oxidation states of these metals. However, for the transition metals the situation is quite different and the relative stabilities of different oxidation states are markedly affected by changes in environment.

Fortunately, using the Nernst equation it is relatively easy to correct for changes in redox potential caused by changes in concentration,

† A simple rule is that the couple with the more positive potential always reacts as written, and the reaction is completed by the other couple reacting in the reverse sense to that written.

‡ Activity is a mathematical device employed in thermodynamics to allow for the fact that solutes of the same absolute concentration may exhibit different *effective* concentrations.

$$E(\text{exp}) = E^{\ominus} + \frac{RT}{nF} \ln \frac{a_{\text{ox}}}{a_{\text{red}}}.$$

If the solutions are dilute then the activities may be replaced by concentrations, so that

$$E(\text{exp}) = E^{\ominus} + \frac{0{\cdot}06}{n} \lg \frac{[\text{Ox}]}{[\text{Red}]}.$$

Inspection quickly shows that a decrease in the concentration of the oxidized species makes the observed potential more negative, whilst a corresponding increase makes it more positive.

It is important to emphasize that changes in concentration are not the only factors affecting the redox potential, though they are undoubtedly important. As has been suggested previously, changes in local environment can have considerable effects, particularly amongst the transition-metal ions. A change in solvent, particularly one involving a significant alteration to the polarity of the medium, is likely to have a much greater effect than even the largest variation in concentration. Similarly there will be a considerable change in redox potential if the water molecules directly bound to the metal ion are replaced by other ligands. Indeed, this latter effect is of considerable importance in regulating the redox potential of many of the transition-metal complexes, especially those containing iron, which function widely as redox reagents for living systems.

Implicit in all the previous discussion has been the assumption that in aqueous solution the metal ion is no longer independent of its surroundings, since due regard must be paid to the coordination of water molecules, as evidenced by the hydration energy. The origin of the hydration energy has been the subject of much discussion, and it is now clear that the problem is quite complex. For the alkali metals at least, the principal contribution to the hydration energy is the electrostatic attraction between the metal ions and the dipolar water molecules. The distribution of valence electrons between the oxygen and hydrogen atoms in water is unequal, and since the oxygen atom is more electronegative (electron-attracting) it carries a partial negative charge, whilst the hydrogens have a partial positive charge. As a result the water molecule acts as a dipole and there is an electrostatic attraction between the oxygen atom and the metal ions. For more highly charged ions such as Mg^{2+} and Al^{3+}, the bonding forces between the water molecules and the cation are certainly more complex. Here the electrons on the oxygen are polarized by interaction with the metal ion to such an extent that the bonding can no longer be considered to be solely electrostatic but must be at least partially covalent, as

Water molecules bound directly to the metal ion form the first hydration sphere

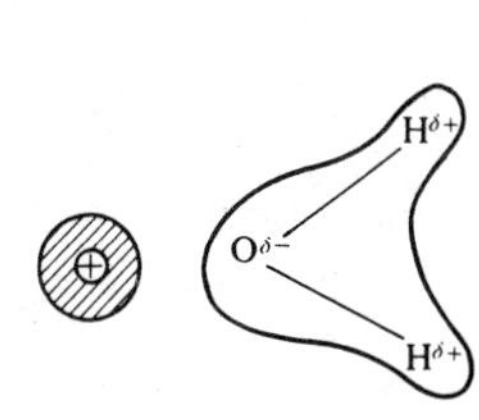

A large ion of low charge gives rise to electrostatic bonding

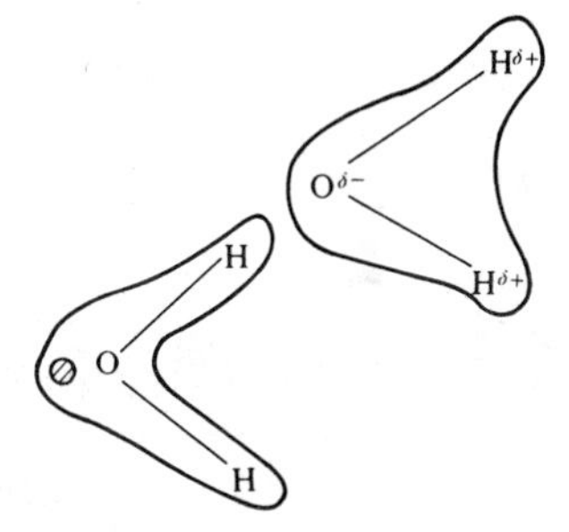

A small, highly charged ion produces considerable polarization and covalent bonding; the hydration sphere is correspondingly extended

FIG. 3. The hydration of a metal ion in solution. The number of water molecules bound to the metal ion depends on both its charge and on its radius. The bonding may be essentially electrostatic or, if the ion is highly polarizing, the bonding may be covalent.

shown in Fig.3. Certainly this is true for the hydrated transition-metal ions. Thus the aquo-ions provide a link between the supposedly simple ionic chemistry of the s- and p-block elements and the coordination chemistry of the d-block transition elements.

It is reasonable to predict that the number of water molecules bound directly to the metal ion – that is to say, those in the first hydration sphere – will depend on both the size and the charge of the metal ion. Experiment and calculation suggest that this number can vary from 4 in the case of Li^+ to 8 or possibly 10 for Ra^{2+}. In principle therefore, it would be logical to assume that the size of the hydrated ion should increase in the same order as the unhydrated ion radius, as derived from crystallographic measurements. However, this presupposes that the effect of the positive charge on the ion is limited to the first coordination sphere. In contradiction, Fig.3 shows that as a water molecule is bound in the first hydration sphere there is a further polarization of its valence electrons which leaves the hydrogens fractionally more positively charged, resulting in a binding of further water molecules to form a second hydration sphere. Such association can then continue throughout the whole system, albeit decreasing rapidly as the distance from the ion increases. This is reflected in the relative magnitudes of hydration energies which generally decrease with increasing free-ion radius. Thus the smaller the ion the greater its polarizing power (as is conveniently measured by the charge/radius ratio) and the larger is the effective hydration sphere. This deduction is confirmed experimentally by measurements of solution viscosity or ionic conductance.

These arguments have been introduced here since they lead to the obvious conclusion that hydration sphere is far from being well defined and will depend on the method of measurement. Taking sodium as an example the measured hydration sphere may vary from as little as six molecules of water, if it is assessed by crystallographic techniques, to hundreds, if mobility measurements are employed. As a result, care must be exercised when using hydrated radii as a basis for explaining cation selectivity in biological systems. In later chapters we shall hope to develop sound arguments in which the general concept of ion hydration will be extended and employed when discussing specific cellular ion selection and rejection processes.

The hydrated metal ion is not an inert unit; even the water molecules in the first hydration sphere exchange with those in the bulk phase of solution with rates that depend not only on the charge and crystallographic radius of the ion but also on its electronic structure. Again the s- and p-block metals exhibit simple behaviour which may be predicted by a simple electrostatic model, whilst, as ever, the transition metals require more detailed treatment for any understanding of their rates of exchange. Nevertheless, such exchanges may be postulated as one of the many factors which allows biochemical differentiation between chemically similar ions such as magnesium and calcium.

Metal ions as Lewis acids

It is convenient to treat the formation of the hydrated metal ion as part of the general process of complex formation between a Lewis acid and a Lewis base. Then all the positively charged metal ions can be classified as Lewis acids, or electron acceptors, whilst the water molecules, which are formally electron donors, can be described as Lewis bases. It is nearly always possible to replace the coordinated water molecules by other Lewis bases, though the molarity of pure water ($H_2O = 55{\cdot}5 \text{ mol dm}^{-3}$) is such that concentration effects in the equilibrium often mask this fact.

$$M^{n+} + xH_2O \longrightarrow [M(OH_2)_x]^{n+}$$

$$\text{Lewis acid} + \text{Lewis base} \longrightarrow \textit{Complex}$$

$$[M(OH_2)_x]^{n+} + \text{ligand} \rightleftarrows [ML]^{n+} + xH_2O.$$

The stability of the resulting metal-ligand complex is further complicated by other factors. First, both acids and bases can be divided into two classes and described as 'hard' and 'soft', mainly on the basis of their polarizability. Non-polarizable metal ions such as Li^+ or Mg^{2+} are 'hard', as are bases with donor atoms such as oxygen, nitrogen, or fluorine, whereas polarizable ions such as Cu^+ and ligands with donor atoms such as phosphorus or sulphur are considered to be 'soft'. It has been observed that the most stable complexes are those formed between either hard acids and hard bases or alternatively between soft acids and soft bases, whilst mixed complexes have lower stability. It is also possible to distinguish between the bonding in the complementary hard–hard or soft–soft complexes. In the first case it is largely electrostatic, and in the second more nearly covalent.

The second factor affecting the stability of a complex is the number of binding sites on the incoming ligand. If the ligand has two or more binding sites which may be used simultaneously (a polydentate ligand) then it can be seen that there will be an overall increase in the number of molecules according to an equation

$$[M(H_2O)_x]^{n+} + L \longrightarrow [ML]^{n+} + x\,H_2O.$$

This may bring about an increase in the entropy of the system and with it a decrease in free energy favouring the reaction in the direction written. This increase in stability is known as the chelate effect, and it is interesting to note that many biological ligands are multidentate.

The geochemistry of the metallic elements

The next step towards understanding the distribution and function of the metals *in vivo* ought perhaps to be a very brief account of their distribution *in vitro*. Certainly, there is no shortage of information for such a discussion. Rather, the opposite is true, and there is now an enormous body of geochemical data available, recording the chemical constitution of the rocks, soils, and natural waters which together form the surface of the earth.

The problem can be attacked in three stages. The first and most fundamental question concerns the origin of the elements, their cosmic abundance, and their distribution. Second is a consideration of the primary distribution of this material in the earth; and third is the review of the subsequent redistribution of the elements during geological history.

The elements themselves are believed to be produced by the fusion of hydrogen by a thermonuclear reaction in the stars, so that the general pattern of abundance of the elements follows a somewhat modified exponential distribution, with the first-formed, light elements having highest abundance (Fig. 4). However, because of the exceptional stability of certain nuclear structures some elements, notably iron and to a lesser extent carbon, oxygen, sodium, and magnesium, have unexpectedly high abundances; whilst others, such as lithium, beryllium, and boron, have lower abundances than predicted, probably because they are consumed in the fusion processes by which heavier elements are formed.

The steps between the formation of the heavier nuclides and the

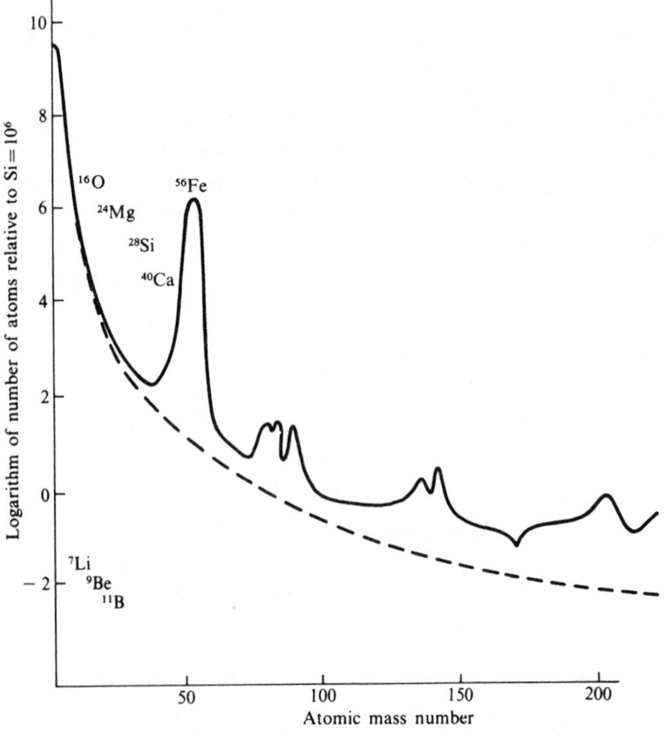

FIG. 4. The abundance of the elements, showing the anomalous abundances and exceptional abundance of iron and the neighbouring elements.

origin of the earth are still obscure, with opposing views suggesting either that the earth was formed by the condensation of hot stellar material or that it grew from the accretion of interstellar debris, followed by a rise in temperature, due principally to radioactive heating. However, it does seem agreed that the molten body which resulted consisted largely of four elements – iron, oxygen, silicon, and magnesium – which, together with a smaller amount of sulphur, are calculated to constitute more than nine-tenths of this planet. The distribution of the remaining elements follows in broad outline the pattern of cosmic abundance, with the exception of the light gases such as hydrogen and helium, which are of lower abundance, having either escaped from, or never having been captured by, the gravitational field of the earth. However, it would be wrong to conclude that the terrestrial distribution of the remaining elements is gravitationally controlled. Instead it may be explained by chemical partitioning, considering particularly equilibria in the iron–magnesium–silicon–oxygen–sulphur system.

In this system the amount of oxygen is very much greater than that of sulphur and the two together are insufficient to combine totally with the remaining three elements. A comparison of thermochemically derived bond energies shows that, of the four remaining elements, silicon has the highest affinity for oxygen, whilst iron has the greatest affinity for sulphur, and moreover is more easily reduced to the metallic state than magnesium. Consequently, the molten earth is believed to have consisted of three essentially immiscible phases – iron, iron sulphide, and magnesium silicate – with the distribution of the minor elements governed by their relative affinities for each phase as shown in Fig. 5. The elements were then divided into several overlapping groups; the siderophile elements, which have a lower affinity for oxygen than iron and which are found in the molten iron core; the chalcophile elements, which have a higher affinity for sulphur than iron, and are also found in the core; and the lithophile elements which are more easily oxidized than iron and so are largely found in the silicate phases of the crust.

As the earth cooled, the elements underwent a second separation, principally by fractional crystallation within the silicate phase of the surface. In this fashion the mantle and crust were formed; in the process dissolved materials such as nitrogen and water were excluded from the solid matrix, resulting in the formation of the atmosphere and oceans. These combined to further modify the surface rocks by the many processes of weathering, producing a variety of sedimentary material, which in turn was compacted to form the sedimentary rocks. In some cases these rocks have been further modified by the effects

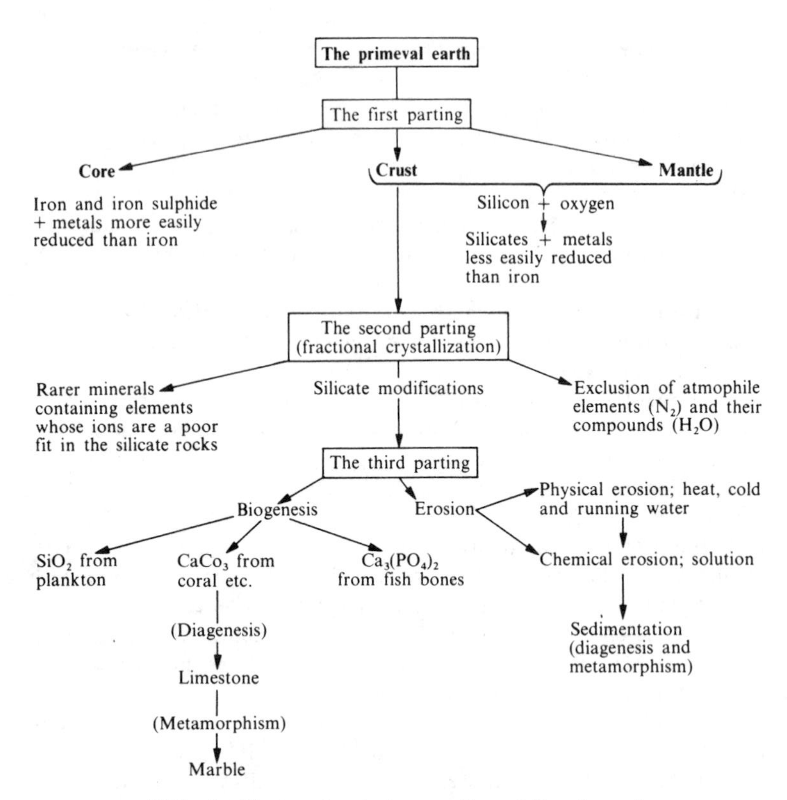

FIG. 5. The geochemical separation of the elements.

of temperature and pressure during geological upheavals, forming what are now known as metamorphic rocks.

Such processes have continued up to the present day, assisted in more recent geological time by biological cycling, including the formation of biogenic deposits such as the carbonate rocks or coal. These processes are summarized in Fig.6. The full importance of biogeochemical cycling is still not understood, and much remains to be learnt about the role of micro- and marine organisms, but it is becoming clear that they can have considerable effect not only in the detailed modification of geochemical distribution but also in the larger-scale processes involved in the formation of bioliths (sediments formed by the residues of living creatures).

The elemental composition of living material

Chemical analysis reveals that most whole organisms are predominantly aqueous, often consisting of more than nine-tenths water,

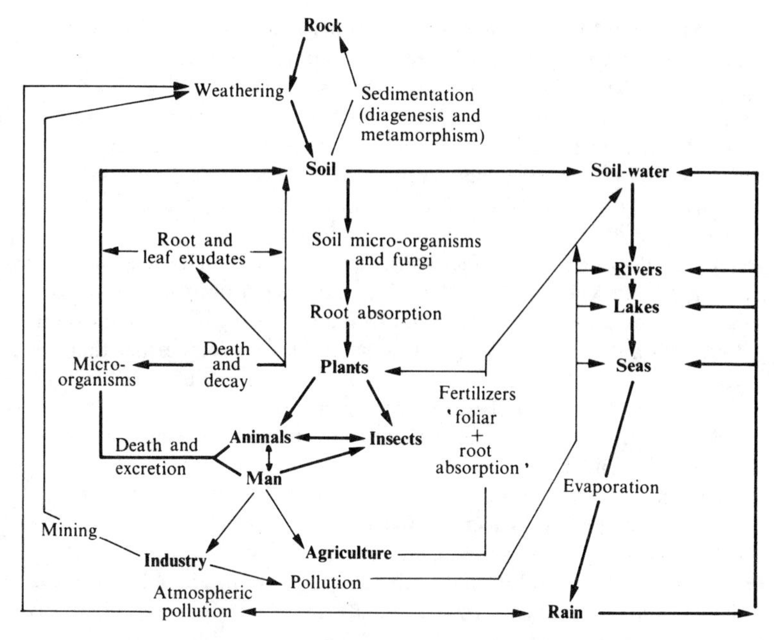

FIG. 6. Some of the major pathways involved in bio-geochemical cycling.

although certain individual organs or structural components may differ markedly from the average composition. An obvious example is the mammalian skeleton which is a highly insoluble polymeric modification of calcium hydroxyphosphate, containing a much smaller percentage of water than the surrounding soft tissue. Apart from any calcareous or siliceous skeletal material, the non-aqueous component of all organisms is largely organic in nature, so that compounds of the metallic elements constitute only a small proportion of the whole organism (the relevant data are listed in later chapters).

Using modern analytical techniques there are no insurmountable problems in determining the absolute amounts of metal ions even at concentration levels of the order of parts per billion (g per billion g†). However, there are a number of pitfalls, both implicit and explicit, in comparing and interpreting the results. Because of the variable water content, it is customary to dry the sample prior to analysis, and then to express the results in convenient units, either parts per billion (p.p.b.) or more commonly parts per million (p.p.m.) of the dry weight. This is usually unexceptionable, except when the initial water content of the sample is uncertain, since results expressed

† A billion = 10^{12}.

on a dry-weight basis have included implicitly a concentration factor ranging from 1 to perhaps 100, corresponding to no water or 99 per cent water in the original sample. Thus results calculated on this basis for the predominantly aqueous soft tissues are artificially large, though correct in relative orders of magnitude. Similarly, many of the metal ions are mobile so that results become meaningless unless it is specified whether the analysis refers to the whole organism, whole organ, or particular cellular or sub-cellular fractions.

Even allowing for such factors the survey of the analytical data now available still reveals that although all the naturally occurring elements of the periodic table may be found in one organism or another, there appears to be only a limited pattern discernible in either their relative concentration or even in the concentration of a single element in different phyla (groups of related organisms).

Examination of the metal-ion concentration of an organism and comparison with what may be termed its growth medium shows that the elements are often present in significantly different concentrations, but the ability to accumulate or reject a specific element is a widely varying phenomenon. These abilities are genetically controlled, but the pattern of evolutionary selection is by no means simple. In the higher plants the capacity to accumulate any single element apparently occurs randomly throughout the different genera, whilst in the lower plants a high concentration of certain elements is characteristic of a particular order.†

Accumulation may reach startlingly high levels. The chromium content of plant ecotypes‡ found on serpentine soils may be over 10^4 times higher than typical levels found in plants growing on other soils.§ Of course accumulation is not restricted to plants, and perhaps two of the best-known examples are the accumulation of mercury in tuna fish and of cadmium in molluscs. In the latter case, concentration factors of over 10^6 are not uncommon when the molluscs are compared with their sea-water environment.

Analysis of whole plants or animals often fails to reveal the true pattern of accumulation, since the absorbed metals are generally not distributed evenly. Consequently, a less obvious but equally important facet of this problem is the ability of certain organs to accumulate metals with respect to the rest of the organism as well as compared with the external background level. A striking example is the concentration of zinc in the retina of the mammalian eye, whilst an obvious but easily

† A genus is a group of animals or plants having common structural characteristics distinct from those of all other groups for example *Homo*, or *Magnolia*.

‡ Variants capable of genetic recombination.

§ Serpentine soil is soil derived from serpentine ($Mg_6 Si_4 O_{11} (OH)_6 H_2O$).

overlooked example is the accumulation of calcium during the formation of skeletal material. Although in both cases these processes are linked with the functional roles of the metals it is certain that neither the presence of an element, nor even the ability to accumulate it against a concentration gradient, are necessarily evidence that it has any function.

Essential, beneficial, and toxic metals

As simple analytical data does not provide a direct guide to the status of a metal, it is necessary instead, in order to be able to describe it as *essential,* to show that its absence will prevent some particular organism from completing its life cycle, reproduction included. In the absence of this element the organism will develop characteristic pathological deficiency symptoms, which may only be alleviated by the particular element acting directly on the nutritional status of the organism and not by influencing the external environment. Using these criteria it has become reasonably certain that in many cases the presence of certain metal ions is merely an artefact and that they are of no functional significance, whilst it has also become increasingly apparent that not all the *essential* elements are necessary to all organisms.

The essential metals listed in Table 1 can be divided into two classes according to the amounts of each which are required to maintain a metabolic balance. Potassium, magnesium, and calcium are generally classified as macro-nutrients, as is sodium for animals, though not for most plants; whilst the remainder are becoming increasingly known as micro-nutrients. This term is replacing the older designation of 'trace element' which occasionally caused some confusion, because some of the so-called trace elements are actually present in quite significant amounts, it is merely the dietary requirement for them that is low.

Whilst it is relatively easy to demonstrate that a dietary insufficiency of the macro-nutrients produces metabolic disorders, the same cannot always be said for the micro-nutrients. The reason soon becomes apparent when it is realized that, in rats, the dietary requirement for chromium, one of the most recent additions to the list of essential elements, is less than 0.1 μg per 100g of body weight daily. Obviously it is extremely difficult to produce a purified and balanced diet with a metal content as low as this, since even the environment of a metal cage can easily give rise to sufficient contamination to provide ample dietary chromium.

Much effort has been expended in overcoming such difficulties. For animals, feeding experiments are now routinely conducted in plastic cages which are themselves contained in a closed environment

TABLE 1

The characteristic properties of the essential elements

There are certain other metals, notably aluminium, nickel, and titanium, whose status is uncertain and many of the remainder are to be found *in vivo* but have no apparent function.

Metal	Chemical characteristics	Biochemical role
Sodium and potassium	The highly mobile unipositive cations form mainly soluble salts but only weak complexes with most ligands with the exception of the cyclic polydentate oxygen donors.	Concentration differences created by active transport provide functional osmotic and electrochemical gradient The ions are also structure-promoters for both polynucleic acids and proteins. Potassium especially is an important enzyme activator by this means.
Magnesium, calcium, and zinc	The divalent ions Mg^{2+} and Ca^{2+} are distinctly hard Lewis acids preferring anionic oxygen donor ligands such as phosphate, whilst Zn^{2+} is borderline and will bind to nitrogen or sulphur donor ligands, especially histidine and cysteine. None of these metals have any redox chemistry.	All the ions act as enzyme activators but this function is least important for Ca^{2+}. They probably act as Lewis acids or structure promoters but *not* as redox catalysts. Magnesium is an integral part of chlorophyll, the photosynthetic pigment in plants. The many insoluble calcium salts and complexes act as structure formers in both plants and animals. The calcium concentration in muscle is actively controlled to act as a neuromuscular 'trigger'.
Vanadium	The most important species are the polyvanadates (which have some similarity to the polyphosphates) and the vanadyl cation VO^{2+}.	Accumulated by ascidian worms and found principally in specific cells (the vanadocytes). The function is uncertain. In mammals vanadium is implicated in cholesterol biosynthesis.

Chromium	Chromium(III) is the principle species. In higher oxidation states its compounds are powerful oxidants.	Chromium(III) is involved in the metabolism of glucose in mammals.
Molybdenum	A congener of chromium but with distinct chemical differences. It has a strong affinity for π-bonding ligands and also a tendency to exhibit high oxidation states.	The exact function of molybdenum is still not clear but it is essential to all nitrogen-fixing organisms as part of the nitrogenase enzyme system.
Manganese	Manganese(II) is quite similar to the alkaline earths but many compounds containing the metal in higher oxidation states are known. These are powerful oxidants.	Manganese(II) acts as a Lewis acid in several enzymes whilst the photosynthetic apparatus requires manganese, apparently as a redox catalyst.
Iron	Iron(II) and iron(III) are the principal oxidation states. Fe^{2+} binds preferentially to nitrogen donor ligands, Fe^{3+} to oxygen donor anions. Cycling between the states is relatively easy and very important.	Iron compounds of two types, iron-protein-sulphide and iron porphyrins are the principal electron carriers in biological redox reactions. The iron-porphyrin-protein, haemoglobin, is the oxygen-carrying pigment of mammalian blood.
Cobalt	Two oxidation states are common: cobalt(II) and cobalt(III); the former is kinetically labile, the latter inert; cobalt(I) may be important biochemically.	The cobalt-corrin complex, Vitamin B_{12}, is an important coenzyme, whilst the Co^{2+} ion is a cofactor for some hydrolytic enzymes.
Copper	The most usual oxidation states are copper(I) and copper(II) and reversible changes between them are not difficult. Copper(III) may be important biochemically.	An important part of many oxidases. It is believed to act by cycling between different oxidation states.

to prevent external pollution, and plants are similarly grown by solution culture in isolated environments. Even so, the status of some elements remains in doubt and will undoubtedly be difficult to clarify.

Another problem is that the animal or plant being studied may have accumulated a sufficient reserve of a particular micro-nutrient so that, despite being fed on unbalanced or deficient diet it fails to show deficiency symptoms; and of course it must be remembered that seeds, spores, or even new-born animals often have considerable reserves of trace elements, sometimes at the expense of the parent. Indeed, in studies on the manganese requirement of rats it has been shown that with animals fed on a manganese-free diet, it may take three generations before true deficiency symptoms are observed.

There is a second group of elements which are widely distributed in living material and which seem to aid growth or reproduction, but whose absence produces no apparent ill-effects. These elements such as lithium or rubidium are often described as *beneficial* elements, particularly in the older literature. Many workers, however, take the view that these will ultimately be proved to be essential, though the requirement may be satisfied by almost vanishingly small amounts.

In contrast there is yet a third group of elements which have neither an essential nor a beneficial but a positively catastrophic effect on normal metabolic processes, even when present in only small amounts. Such elements are commonly described as toxic, though the term itself is only loosely defined. The pathological effects and significant dose in metal-ion poisoning are remarkably complex and variable, depending particularly on those factors that can modify the uptake of the metal as well as those which control its subsequent metabolism, and many of the problems which have to be resolved in establishing the lethal dose are the same as those related to determining the minimum requirement for an essential metal.

The process of evolutionary selection has resulted in many organisms developing tolerances to what are normally considered toxic elements. It is well established that such tolerances are element-specific. Copper-tolerant plant ecotypes, for instance, usually exhibit no significant increase in tolerance towards nickel or zinc, the neighbouring elements (this is used as the basis for geo-botanical prospecting). Well-known examples are the serpentine, seleniferous, and uraniferous floras. In all cases the boundary between the indicator species and the surrounding vegetation is sharp and the range of the tolerant ecotype may be very limited. On localized deposits such as mine tailings, the range of the tolerant species may be just a few hundred square yards. The specific mechanism of tolerance is as widely varied as its occurrence though it can be generalized into rejection mechanisms and protective metabolizing which, surprisingly, may include accumulation followed by immobilization.

From a large number of simple feeding experiments it is now becoming clear just which elements are essential, which are toxic, and which have neither specific function nor particularly harmful effects. However, what also emerges from these studies is that it is unwise to consider the effect of each element in isolation, though it is still not possible to give a detailed account of their many interactions. More and more it becomes obvious that the essential metals, even the micro-nutrients, do not operate in well-separated metabolic pathways (if indeed such unlikely phenomena exist) but that they all interact to varying degrees. For example, it is certain that in mammalian iron metabolism the incorporation of iron into the respiratory pigments is directly and substantially affected by both copper and cobalt compounds, whilst to a lesser extent it is also modified by changes in the metabolism of molybdenum or zinc which act indirectly through copper. Since numerous other interactions modifying zinc or molybdenum metabolism are known it is soon made clear that changes in the balance of any one element will affect all the others, either by direct antagonism or indirectly through secondary metabolic changes.

Over all, the problem of how and why a particular metal affects a certain organism can be attacked in three stages: the first is an appreciation of the chemical, physico-chemical, and biological factors affecting initial availability; the second is the factors affecting absorption, translocation and retention; and the third is the investigation of the factors affecting utilization.

The availability of an element is not necessarily the same as its natural abundance in the earth's crust. Nickel, for example, appears to have only a minor role even though it has the same natural abundance as zinc, and zinc is certainly an important essential element. This may well be due to the fact that nickel, unlike zinc, is strongly absorbed by the silicate rocks and soils and hence is not readily available. On the other hand, it has long been known that even relatively minor changes in the environment can cause drastic changes in the availability of a metal ion. In agriculture, for instance, it is common knowledge that excessive liming can be harmful. In this case the cause is clear: the addition of lime immobilizes the heavy metals by their precipitation as hydroxides.

The absorption of metal ions cannot be described by any single model. The rate of absorption depends critically on chemical state, thus in man the absorption of iron increases on transforming the iron from the metallic state to simple salts and then to complexes with naturally occurring ligands such as citrate, but the complications are enormous. Salts of iron(II) are absorbed more rapidly than those of iron(III), whilst positively charged complexes such as tris-(2,2 -bipyridyl) iron(II) are much less easily taken up than negatively charged complexes containing iron in the same oxidation state. Similar comments can be made about the metabolism of almost all the other metals though the amount of experimental information which is available varies widely. What has become apparent is that the absorption of essential micro-nutrient elements is often mediated by specific mechanisms which are usually linked with subsequent translocation and storage processes. Examples, where known are quoted in later chapters.

The utilization of a metal may also depend on its chemical form. Thus pernicious anaemia, which results from a deficiency of Vitamin B_{12}, (a cobalt–corrin complex) is alleviated only by supplementation of the diet with the intact cobalt complex; simple cobalt salts or other cobalt complexes are mildly toxic. In contrast, iron-deficiency anaemia will respond to treatment with a wide variety of iron compounds, with the proviso that the efficacy of such treatment will depend on the rate of uptake.

2. The alkali metals

The chemistry of the alkali metals

THE alkali metals,Group IA, have probably the simplest chemistry of any group of elements in the periodic table. The metals themselves are stable only under anhydrous conditions since they all react with water to form their hydroxides. The reactivity of the metals increases on going down the group, reflecting the decrease in ionization energy with the increases in atomic radius.

In each case the first ionization energy is low, so that the necessary energy is readily provided by the hydration of the product ion and as a consequence the relevant redox potentials are all both large and negative, as shown in Table 2. (At first sight it seems curious that despite the variation in ionization energy the associated redox potentials are almost constant, but this is merely fortuitous, as the decrease in ionization energy just happens to be paralleled by a decrease in the hydration energy.) In every case the loss of a further

TABLE 2
Some properties of the alkali metals

	Li	Na	K	Rb	Cs
Electronic configuration	(He) $2s^1$	(Ne) $3s^1$	(Ar) $4s^1$	(Kr) $5s^1$	(Xe) $6s^1$
Ionization potentials ($kJ\ mol^{-1}$)	517 7263	493 4540	416 3054	401 2650	373 2270
Enthalpy of hydration ($kJ\ mol^{-1}$)	-515	-406	-322	-293	-264
$E^{\ominus}(M^+/M)$ V	-3·02	-2·71	-2·92	-2·99	-3·02
Ionic radius (nm)	0·060	0·095	0·133	0·148	0·174
Ionic mobility† ($m^2\ s^{-1}\ V^{-1} \times 10^8$)	4·0	5·2	7·6	8·1	8·1

† Unfortunately many texts mistakenly report data for ion conductivities in place of mobilities. The conversion is simple:

$$\text{mobility} = \frac{\text{ionic conductivity } (\Omega^{-1}\ m^2\ mol^{-1})}{F}.$$

electron is precluded as the second ionization potentials far outweigh the possible hydration energies of the +2 cations. The difference in magnitude between the first and second ionization potentials can easily be understood in terms of the electronic configuration of the elements concerned. These are also shown in Table 2, together with other representative data. In each case the neutral atom has a single s-electron outside a core of closed-shell configuration. As these inner electrons form an effective shield around the nucleus it requires relatively little energy to remove the outermost s-electron. In contrast, the second ionization potentials are all huge since the removal of more than one electron means that the particularly stable closed-shell configuration already achieved by the unipositive ion must necessarily be disturbed. Consequently, in any aqueous biological system it is solely the chemistry of the univalent cations which needs to be considered in detail.

The loss of the outermost s-electron has two important results, namely, that the product ion is spherically symmetrical and also that the remaining electrons are held tightly by the excess positive charge. This means that the ions are non-polarizable and consequently can be treated quite simply as charged spheres, using the mathematics of classical electrostatics, so that the charge/radius ratio provides a useful measure with which to rationalize trends within the group. It also provides a means of comparing these unipositive ions with the more highly charged ions of other groups. Thus the many similarities between the chemistry of lithium and magnesium can be explained by the similarity in the charge/radius ratio of their ions, Li^+ and Mg^{2+}

Perhaps the most important trend within the group which is satisfactorily explained on this basis is the variation in hydration energies of the ions. Lithium, with the highest charge/radius ratio, has the strongest interaction with the oxygen atom in water, that is, the negative end of the water dipole, and hence has the largest hydration energy, whereas caesium, the largest ion, has the lowest. This in turn explains the apparent anomalies in the ionic mobilities. Caesium, which has the largest † 'naked' or crystallographic radius and might therefore be expected to have the lowest mobility in solution, has in fact the largest. Instead, it is the hydrated lithium ion, whose *effective* radius is largest, which has the lowest mobility. The importance of ion hydration cannot be over-emphasized as it plays such a fundamental part in generating the ion selectivity which is so vital a characteristic of biological membranes.

The most common compounds of the alkali metals are their salts,

† Neglecting francium, which for most biochemical purposes may be safely ignored.

which are formed by the reaction of either the hydroxides or of the metals themselves with acids. It is impossible, of course, to define a molecule of such a compound since even the crystal consists solely of discrete ions held together in a regular lattice, primarily by electrostatic attractions. In general, the intrinsic energy of such an array, the lattice energy, is highest when all the ions are of similar size so that they pack most closely together. It is the balance between the lattice and hydration energies which controls the solubility.

Thus, in the case of large anions such as perchlorate, the lithium salt should be most soluble, for not only will the lattice energy be small, due to the large difference in ionic sizes, but also the hydration energy of lithium is the largest of all the alkali-metal cations. On the other hand, when small anions like fluoride are considered, the problem is more difficult since the lithium salt will now have the highest lattice energy whilst still having the largest hydration energy. Indeed, only small changes in the balance between these two energies induce considerable changes in the solubility. So, for example, potassium nitrate is soluble, yet the standard free energy of solution of the insoluble potassium perchlorate is only 12 kJ mol^{-1} more positive (less favourable).

Not surprisingly, the alkali-metal salts of simple mineral acids are generally much less soluble in liquids which are less effective at ion solvation than water, and in non-polar solvents such as the hydrocarbons the simple alkali-metal salts are totally insoluble.

In most cases, then, the alkali-metal salts are relatively soluble in water so that the solid salts are biologically unimportant and it is the properties of the ionic solutions which are of fundamental interest. Unfortunately, no adequate theory exists which is capable of dealing comprehensively with such systems,† though it is still possible to arrive at a number of experimentally verifiable conclusions with the aid of the simplest models and the minimum of mathematical manipulation.

Once in solution, the ions will be separated and free to move about. However, despite the 'insulation' of their solvent spheres, oppositely charged ions will still attract each other, forming ion-pairs. Even in a dilute solution of lithium nitrate the ion-pairs are about one-tenth as abundant as the free ions and in solutions of the sulphate salt this rises to over six-tenths by virtue of the higher charge of this anion. Of course, such ion pairs are constantly being broken up and re-formed and this means that it is often possible to consider the cations quite separately from the anions, apart from the proviso that overall charge

† It has been unkindly said that the Debye–Hückel treatment, which is most commonly used, is applicable only to slightly contaminated distilled water.

neutrality must be maintained. This has important consequences for biological systems and, as will be seen, the functional activity of sodium and potassium can often be discussed without specifying the nature of the counter-ions.

It has already been pointed out that the aquo-ion can be regarded as just one example, albeit an important one, of a complex between a Lewis acid and a Lewis base. In principle there is no reason why any other anion or dipolar molecule should not be able to form a similar complex with any of the alkali-metal ions, though of course the metals, which are all 'hard', will have little affinity for ligands with

D-hydroxyisovalerate
L-valine
D-valine
L-lactate

Valinomycin, a cyclic pseudo-peptide antibiotic

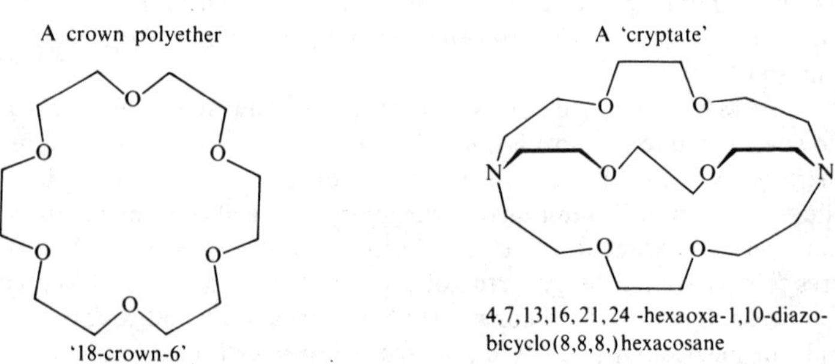

FIG. 7. Some cyclic chelating ligands. The selectivity of ionophoric ligands can be controlled by altering the ring size.

'soft' donor atoms, and will prefer 'hard' donor ligands, particularly those which bind through oxygen. The bonding in such complexes will be largely electrostatic so that simple monodentate ligands: ethers, alcohols, phenols, esters, etc. will not easily displace the coordinated water molecule mainly because of its higher effective dipole strength. However, the situation is greatly altered if the incoming ligand is polydentate, and most of the known complexes of the alkali-metal cations contain this type of ligand.

Until recently the majority of complexes were formed from β-diketonate or phenolate ligands. The properties of these complexes readily distinguish them from the simple salts. Sodium benzoylacetonate dihydrate for example is soluble in toluene, which is not true for any simple salt. More recently three new classes of multidentate ligands have come into prominence. These are the cyclic polyethers, the nitrogen–oxygen macrocycles (the so-called 'cryptates'), and the naturally occurring cyclic peptide antibiotics, examples of which are shown in Fig. 7.

Since chelation is necessary in order for ion-binding to occur, all the polar groups of the ligand point inwards, leaving the apolar side chains on the outside. This gives the complexes great solubility in non-polar environments, so much so that some polyethers are able to sequester the alkali ions from aqueous solutions to non-polar solvents such as chloroform or even ether. In biological systems, the cyclic peptides are effectively able to 'hide' the metal ion and 'smuggle' it through the non-polar lipid layer of the cell membrane, bypassing the normal transport processes and radically altering the pattern of ion distribution, which no doubt accounts for their antibiotic activity. Clearly, ion binding by such ligands and the increased lipid solubility it confers will prove to be of increasing importance in understanding the translocation of sodium and potassium in cellular membranes.

Methods for studying the alkali metals *in vivo*

In many ways the simplicity of the chemistry of the alkali metals makes them a difficult group of elements to study. Conventional 'wet' chemical analysis rests on the formation of insoluble or coloured salts and complexes, or on redox reactions. Clearly, the latter can have no application in this case, nor unfortunately do the alkali metals form many insoluble compounds suitable for gravimetric analysis. The mixed 'sodium zinc uranyl acetate' and the tetraphenylborates are amongst the few that are used at all commonly. Colorimetric methods of estimations are equally rare and of course as the alkali-metal ions have a closed-shell configuration they have no low-energy electronic transitions and so exhibit no simple room-temperature electronic absorption spectra. However, by heating the ions in a flame they may be made to emit rather than to absorb light and this is the basis of the most widely used analytical method for these elements, flame emission spectrophotometry (flame photometry).

Using this technique the total analytical concentration of the alkalis can be measured quite easily and satisfactorily, though in view of their mobility the results must always be interpreted with care. For example, analysis of whole animal tissue would not reveal the concentration differences between the plasma and the intracellular fluids, nor would it reveal that such differences are maintained by constant ion fluxes across the cell membrane. Such details of the metabolism have largely been uncovered by the use of tracers.

Lithium has no suitable isotopes for tracer work, whereas sodium has two, ^{22}Na, (a positron emitter; half life of 2·6 years) and ^{24}Na (a β^- emitter; half-life 15h). Generally ^{24}Na is used since it is cheaper, being readily prepared by neutron irradation of ^{23}Na, but in studies requiring months rather than days, ^{22}Na is the isotope of choice. The radioactive isotopes of potassium are less useful, either by virtue of their low natural abundance and long half-life, for example, ^{40}K (0·01 per cent abundance; β^- emitter; half-life $1{\cdot}2 \times 10^9$ years) or expense, for example ^{42}K (β^- emitter; half-life 12·5 h). Geochemical dating using the $^{40}K : {}^{40}Ar$ ratio is however of considerable importance.

However, *in vivo* these difficulties may be partly overcome by using rubidium as a probe for potassium. Many studies of potassium distribution have been carried out using ^{87}Rb, (27·85 per cent abundance; β^- emitter; half-life 6×10^{10} years) employing the principle of 'isomorphous replacement'. This originated in geochemical studies where it was noted that ions of similar radius (though not necessarily of equal charge) often substituted for each other in otherwise identical lattices. The principle has since been extended to encompass living systems, though not without problems caused by the subtle nature of biological discrimination. Apart from rubidium, both ammonium and thallium (I) may substitute for potassium and the latter is particularly useful as it possesses both a readily observable electronic absorption spectrum and significant paramagnetism. Isomorphous replacement has been mainly restricted to studies on potassium metabolism as there is no really good probe for sodium, lithium being too small and potassium too large.

The absence of unpaired electrons in all the alkali-metal cations means that they cannot be studied by the magnetic measurements which are so important for the transition metals, and neither bulk-susceptibility measurements nor electron spin resonance (e.s.r.) spectroscopy are useful. However, the absence of paramagnetism does mean that is is possible to study the ions directly by means of nuclear magnetic resonance spectrometry. Both ^{7}Li (92·7 per cent abundance) and ^{23}Na (100 per cent abundance) have non-zero nuclear spin, I = ½ in both cases, and so give rise to easily observed spectra. As yet relatively little work has been done in this field but it promises to be of increasing importance, since in principle, if not yet in practice, this technique may give information about the chemical environment of the metal ions in intact samples.

The terrestrial distribution of the alkali metals

The six elements of Group IA, all lithophiles, can be found in widely varying proportions distributed through the complex alumino-silicates which make up so much of the earth's surface. In such minerals the ratio of oxygen to silicon varies considerably, but is always greater than 2:1, so that there is a large series of silicate lattices containing various anionic sites. Some of the sites in each silicate cage are filled with Al^{3+} ions, and the size and shape of the remaining 'holes' generate a preference for one particular alkali-metal cation.

The least abundant is the heaviest member, francium, which is found only in minimal traces and may be largely ignored. All of its isotopes are radioactive and even the longest-lived ^{223}Fr, formed by the decay of ^{227}Ac, itself decays to ^{223}Ra, with a half-life of only 22 min.

Interestingly, lithium is generally not associated with the remaining members but instead is more commonly found in the ferromagnesium minerals, where it may partly replace magnesium. The average concentration of lithium is relatively low, 60 p.p.m. or less, whilst that of the next member of the group, sodium, is both much greater and also much more variable, with concentrations from 400 p.p.m. in limestone to nearly 30 000 p.p.m. in igneous rocks. Again, sodium is not usually found in conjunction with either lithium or potassium since the differences in ionic radii, Li^{+} = 0·060 nm, Na^{+} = 0·095 nm, K^{+} = 0·133 nm, are too marked. However, the remaining three members of the group (K^{+}, Rb^{+}, Cs^{+}) do occur almost interchangeably, with potassium by far the most abundant.

Much is now known about the geochemical cycling of these elements, particularly about sodium and potassium. The leaching action of rain on rocks and soil has gradually washed out some of these ions, mainly sodium, which is generally held least strongly, with the result that the sea is now quite strongly saline. The large oceans are now about 0·46 mole dm^{-3} NaCl, roughly 1 per cent by weight, and of course this will increase, albeit slowly, in the future. This process is accelerated in enclosed waters such as the Dead Sea, which suffer a high rate of evaporation, and in the geological past the complete evaporation of primeval saline lakes had led to the formation of extensive salt pans. Although the general trend in movement of dissolved salts is into the sea there is some transport back to the land via wind-borne water and the so-called 'cyclic salt' has been detected in water droplets at altitudes up of to 2000 m..

The alkali metals *in vivo*

All the alkali metals (excepting francium) are widely distributed in living material, but only sodium and potassium have well-characterized functional roles and are considered to be essential. Lithium, rubidium, and caesium are found in most organisms at levels higher than in the background environment, but nevertheless at lower levels than sodium and potassium. Experiments show that each is able to mimic at least some of the functions of one of the essential elements, but it is still not certain whether this implies any strict functional role. Currently, lithium appears to be the most important since some plants such as tobacco only grow poorly without it. It also has a distinct effect on the nervous system of the higher animals, probably by partly replacing sodium. Lithium carbonate is used in the treatment of manic-depressives, but the dose must be stringently controlled since, at plasma concentrations greater than about 1×10^{-3} mole dm^{-3} it rapidly produces *diabetes insipidus,*† particularly in patients with hyponatraemia.‡

When discussing the distribution and function of sodium and potassium it is convenient to distinguish between plants and animals. Potassium is abundant and essential in both classes, whereas sodium is largely restricted to animals. Indeed, many plants will not tolerate even mildly saline conditions, though equally it must be pointed out that there are a number of strict halophytes§ which require salt, though the requirement is probably more for chloride ion than for sodium.

From the earliest recorded times salt has been used as an article of currency and it is no accident that either to offer or to accept is is a common way of expressing friendship or suspending hostilities. Whilst most people realize its importance there are peculiar exceptions. At least one tribe in South America is reported to have no readily available source of salt and so uses none in its cooking. Consequently, they suffer from the typical symptoms of salt deficiency: dehydration, loss of coordination, and poor reflexes. Estimates of the amount of salt

† Symptoms include chronic excretion of urine and extensive thirst, probably caused by interference with kidney function (see p. 34).

‡ Hypo- is a prefix meaning 'below normal'. Thus hyponatraemia is a condition in which the sodium level is below normal.

§ Salt-requiring plants.

necessary for a balanced diet in man vary considerably but commonly accepted figures vary between 1·0 and 2·0g per day and a considerable proportion of this is supplied in the cooking process. Salt balance in land animals is generally controlled by urine formation and hence is inseparable from the water balance. Replacement of lost salt is vital since without it it is impossible for the body to absorb the necessary water to make good any losses. What is not commonly realized, however, is that sodium chloride alone is relatively ineffective and a more effective treatment for patients who are severely dehydrated involves judicious medication with both sodium and potassium salts. Just as serious as salt deficiency is the opposite, hypernatraemia†, though it is more rarely met in practice. The emetic effect of salt-water is well known, but lower doses over a longer period result in a much more

† Strictly this refers to higher-than-average levels of sodium in the plasma

TABLE 3

The approximate concentrations (in p.p.m.) of sodium and potassium in various biological materials

	Sodium	Potassium	Ratio†
Rock and soil‡ : range	400 - 25 000	2000 - 30 000	-
Fresh-water	10	5	-
Sea-water	10 500	380	47 : 1
Mxyine (whole blood)	12 500	350	61 : 1
Crab (whole blood)	11 000	470	40 : 1
Crab (nerve)	11 300	4500	0·5 : 1
Squid (whole blood)	10 000	750	23 : 1
Squid (nerve : giant axon) §	1100	15 600	0·12 : 1
Arbacia (eggs) §	1200	7500	0·25 : 1
Valonia §	1900	19 500	0·17 : 1
Nitella §	250	2200	0·19 : 1
E. Coli §	1800	8500	0·35 : 1
Camel (whole blood)	3500	200	30 : 1
Rat (whole blood)	3200	250	22 : 1
Man (whole blood)	2000	1350	2·5 : 1
(erythrocyte) §	150	6000	0·04 : 1
(plasma)	3500	150	38 : 1
(muscle) §	600	4000	0·25 : 1

† Calculated as relative *numbers* of atoms (corrected for relative atomic masses).
‡ Expressed on dry-weight basis.
§ Expressed as p.p.m. of intact wet tissue.

dangerous condition characterized by the retention in the body of excess water.

In the healthy animal the need for potassium is easily satisfied by the normal diet without recourse to supplement, since both plant and animal cells are relatively rich in potassium. For this reason it is only more recently that the essential nature of potassium salts for animals has been recognized, though it has long been known to be necessary for plants, and potassium nitrate forms an important part of many fertilizer formulations. Potassium deficiency is not common in animals though in man it is often artifically and accidentally induced by taking diuretics. This is a particular problem in the clinical treatment of certain cardiac conditions and has led to the development of 'slow-K', a slow-release potassium supplement.

Some typical concentrations of the alkali metals in living matter are given in Table 3.

The functions of sodium and potassium

Despite the high mobility of both sodium and potassium they are not distributed entirely passively so as to minimize osmotic or electrochemical gradients. Indeed, just the opposite is true, and energy is utilized in the active control of the cation concentrations in order to create osmotic or electrochemical differences.

In animals sodium as sodium chloride is the principal solute of the extracellular fluids, and as such is largely responsible for the osmotic balance of the body as a whole and hence the tissue water content. Sodium ions play a part in the maintenance of the acid–base balance of the body and they are also involved in the production of the electrical impulse of the nervous system. More recently it has become apparent that sodium has a number of other functions which do not depend on the creation of concentration gradients. Because of the experimental difficulties inherent in studying the formation of alkali-metal-ion complexes it has only just become apparent that the sodium ion may play an important role in both nucleic acid and protein chemistry, since by binding at anionic sites on the macromolecules it may help to stabilize a particular conformation.

Potassium is essential to both plants and animals and though some of its roles are similar to those of sodium, the two ions are not interchangeable. In both plants and animals potassium is now well known to be an important enzyme activator. It is believed to operate like sodium by binding to anionic sites and thus controlling the protein conformation. In accordance with the known instability and lability of alkali-metal complexes potassium concentrations of $50-100 \times 10^{+3}$ mol dm^{-3} are commonly required to give maximum activation. In a few cases direct experimental verification has been obtained. Thus with aspartokinase–homoserine dehydrogenase the large changes which can be induced in the protein absorption spectrum on binding substrate or inhibitor molecules are apparent only in the presence of potassium ions, sodium ions being ineffective. There is no doubt that proteins or polynucleic acid molecules can readily distinguish between the two ions, but unfortunately, few quantitative data are available. Potassium has also been implicated in the mechanism of cellular replication, probably by controlling the RNA conformation, and certainly *in vitro* experiments show that distinct spectroscopic changes occur in the nucleic acid fluorescence spectrum when alkali-metal ions are introduced.

Potassium also has a number of other functions which really depend on the creation of concentration gradients. These include the creation of transmembrane potentials and the secretion of gastric acid.

Sodium and osmo-regulation

A low intracellular concentration of sodium is typical of a wide variety of organisms (Table 3), and it seems probable that the ability to reject sodium was developed early in the evolutionary time-scale as one method of controlling the cellular water-balance. Only the hagfishes (*Myxine*) have blood which is similar in ionic concentration to the sea, yet most marine organisms have membranes which are permeable to water but impermeable to charged solutes, particularly inorganic cations. This is particularly true of fish gills which must be thin-walled to allow the exchange of gases. Therefore, since the intra-cellular fluid contains a wide variety of organic solutes in addition to the inorganic ions it should clearly absorb large quantities of water by simple osmosis. In order to oppose this the total osmolar concentration of the intracellular fluid is reduced by the rejection of sodium ions.

The choice of sodium rather than potassium is not arbitrary, but has been dictated by their relative concentrations in sea-water. The external concentration of potassium is low so that the maximum concentration difference which could be obtained even by the total exclusion of potassium ions is small. A much larger and hence osmotically more significant difference is obtained by excluding sodium ions. Studies on the egg of the sea-urchin (*Arbacia*) have shown that the removal of the sodium from the cell occurs continuously against the concentration gradient and must therefore require energy. This is supplied by the hydrolysis of adenosine triphosphate (ATP).

Osmo-regulation in *Arbacia* eggs is fairly rudimentary since they cannot cope with changes in external environment. Placed in hypertonic saline† the eggs shrink due to loss of water whilst in hypotonic solutions they absorb water and expand, and in distilled water they lyse.‡ In contrast the eggs of fresh-water fish have solved this problem by becoming impermeable to water. The adults achieve balance in a different fashion; they excrete a very dilute urine. Some marine species, notably the teleosts, have blood which is actually hypotonic with respect to the sea; these maintain balance by pumping out excess salt through special glands in the gills. Similar mechanisms are found in sea birds, which have nasal salt-glands for just this purpose, and even in the plant kingdom, in certain species which grow in salt-marshes.

In man and other animals the water balance is maintained mainly by the production of urine, and sweating is of relatively minor importance

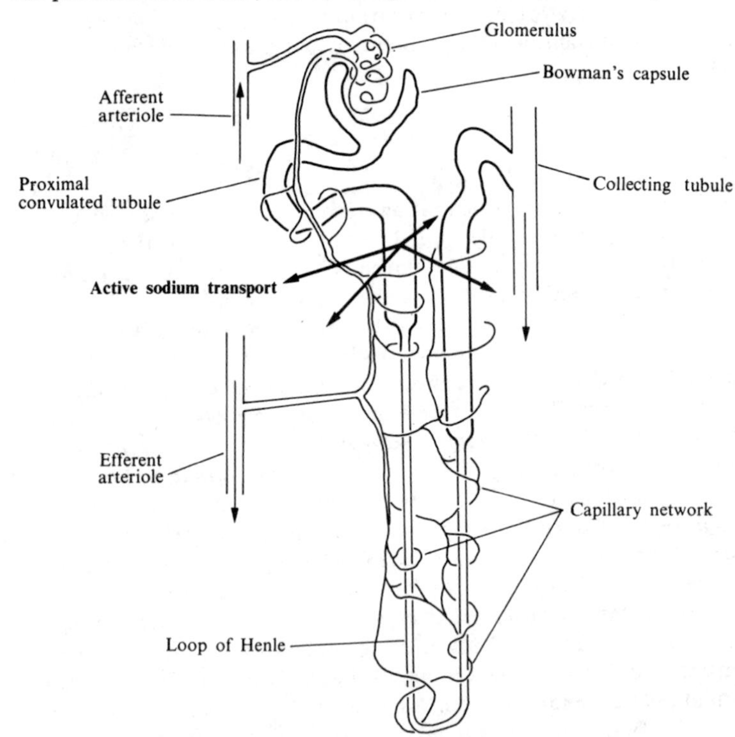

FIG. 8. The nephron. The relative length of the loop of Henle correlates well with habitat, being short in aquatic or semi-aquatic mammals and long in those living in arid or desert environments.

† Isotonic solutions have the same osmotic pressure.

‡ Burst due to internal pressure of water absorbed to relieve excess internal osmotic pressure.

for this purpose. Urine formation occurs in the kidneys and each kidney consists of over a million morphologically and functionally similar units, each known individually as a nephron (Fig.8). The process of urine formation begins by passage of blood through the capillaries of the glomerulus which effects an ultrafiltration so that large molecules are not accidently lost. The rate of glomerular filtration is approximately 125 cm^3 min^{-1} which means that a total volume of about 180 dm^3 is filtered each day, yet the average output of urine is between 1 and 2 dm^3. Clearly, therefore, the majority of the glomerular ultrafiltrate must be reabsorbed.

Resorption occurs primarily in the proximal tubule, where both sodium and potassium ions are *actively* transported across the epithelial cells and eventually back into the plasma.† Resorption of the chloride ion follows *passively* down the electrochemical gradient set up by movement of the cations, and water is similarly resorbed down the osmotic gradient. This is why a patient whose body has been depleted of both salt and water is not able to replace the fluid without salt.

Sodium transport and acid–base balance

Acid–base balance in the plasma is maintained by the buffering equilibrium between carbon dioxide, as carbonic acid, and the hydrogen carbonate anion (HCO_3^-). The kidney is involved in the process by regulating the concentration of HCO_3^- ion in several ways. Circulating HCO_3^- appears in the glomerular filtrate but is *indirectly* resorbed via the transfer of hydrogen ions. Fig. 9 shows a simplified scheme for this process. In a similar way sodium ions are involved in restoring the hydrogen carbonate reserves in the kidney which are depleted in buffering the sulphuric and phosphoric acids produced in various metabolic pathways.

The molecular mechanism of active transport

In *in vitro* experiments using erythrocytes (red cells of the blood), it is possible to change the concentration of sodium ions in the bathing fluid without significantly altering the intracellular ionic concentrations. This implies that the membrane is impermeable to sodium ions. Nevertheless, if ^{24}Na is introduced into the bathing medium then after a fairly short interval it can also be found inside the erythrocytes. It can be concluded, therefore, that the membrane is in fact permeable to sodium ions, and consequently the low intracellular sodium concentration must be the result of a *dynamic*, rather than a static equilibrium.

† The epithelium is the purely cellular avascular layer of the structure.

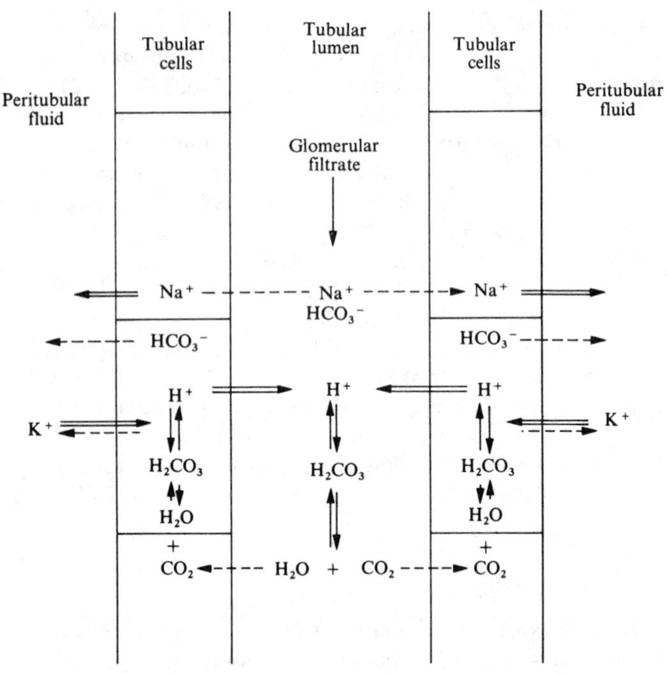

FIG. 9. The renal regulation of acid-base balance. In order to maintain the ionic balance, the movement of hydrogen ions is counteracted by the active transport of sodium ions in the opposite direction.

Further evidence is supplied by treating erythrocytes with metabolic inhibitors, such as cyanide or fluoride ion, alcohol, or the cardiac glycoside, oubain, whereupon the ionic gradients diminish until passive osmotic equilibrium is reached. A similar effect can be obtained by cooling the cell to slow down the rate of metabolism, but in this case the effect is reversible and by careful warming in the presence of an energy source such as glucose, the ionic gradients are re-established. Taken in conjunction, these experiments provide convincing evidence that the ionic gradients are maintained by an active, or energy-requiring process, generally known as the 'sodium pump', and numerous experiments have now demonstrated quite clearly that the hydrolysis of ATP to adenosine diphosphate (ADP) and inorganic phosphates is the energy source of the transport process, though at one time it was thought that the electron-transport chain might be involved instead.

The sodium pump might equally be called the potassium pump as experiments with tracers have shown that in the human erythrocyte three sodium ions are carried out of the cell and two potassium ions moved in, for each molecule of ATP hydrolysed. In other cells, the stoichiometry may be different, thus in the giant axon of the squid, for example, the ratio is $3K^+$ in : $3Na^+$ out : 1 ATP.

Nevertheless, the cell membranes and microsomes of all cells which exhibit active transport have been shown to contain rather similar ATP-hydrolysing enzyme systems. Extensive studies on such systems have all been hindered by the fact that the ATPase complex is firmly bound to the cell membrane, but it has been shown that at least two enzymes are present. The first is responsible for the hydrolysis of ATP and requires both sodium and magnesium ions for maximum activity. This produces ADP and the phosphorylated form of the actual carrier,

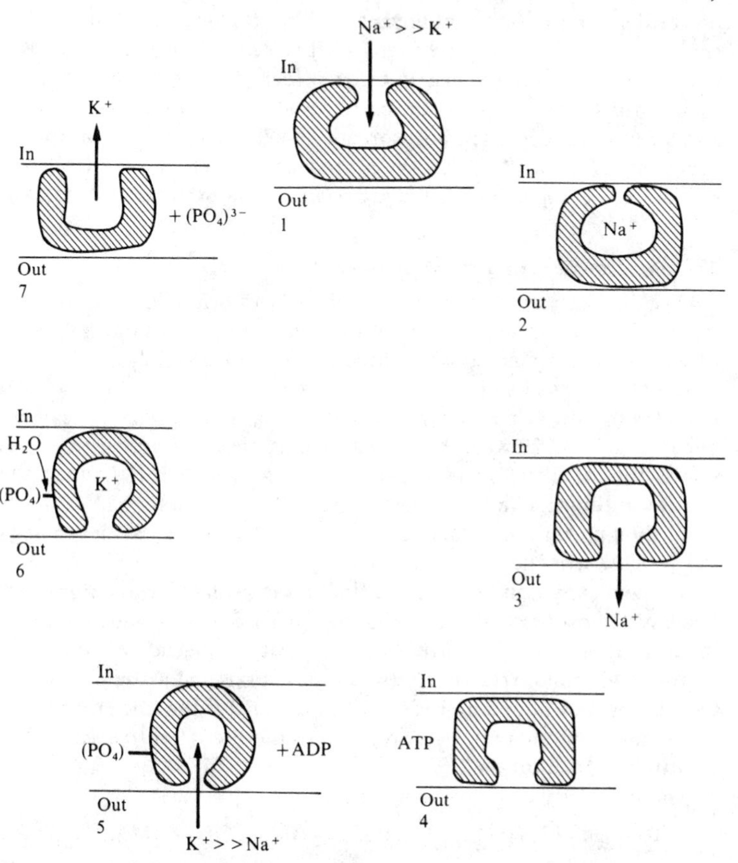

FIG.10. A schematic representation of active transport. Inside the cell the carrier preferentially binds sodium ions, causing a change in conformation which ejects the ion from the cell. The complementary process of potassium transport is brought about by phosphorylation of the carrier, using ATP as the source of energy.

which itself is now thought to be mainly protein in nature. The second enzyme, which requires potassium ions, is then responsible for the dephosphorylation of the carrier and the final production of inorganic phosphate. Fig.10 shows a general scheme which accounts for these and many other experimental observations.

This scheme cannot be regarded as in any way definitive, since at least two problems of particular importance remain unsolved. It has still not been established whether there are two sets of sites which allow synchronous transport in opposite directions, or whether the carrier has only a single set of sites, which alternately carry sodium ions outward and potassium ions inward. Nor has the mechanism of specificity in ion binding been settled. It is clear, however, that phosphorylation provides the energy for translocation, and it well may be that, in the process, it alters the carrier-protein conformation in such a way as to produce discrimination and specific ion binding. On the other hand, it is possible that the protein conformation is controlled by the ionic composition of the intracellular and extracellular media.

The physico-chemical basis of ion selection

The discrimination between sodium and potassium ions exhibited by the active transport system is only a part of the general problem of discrimination amongst the whole series of alkali-metal ions by different biological membranes. At first sight this appears to be a complex problem since the five elements (neglecting francium) can be selected in 5! = 120 ways, but it is a striking observation that the selectivities of all cell membranes, along with most artificial membranes, peptide antibiotics, ion-exchange resins, natural minerals, and artificial alumino-silicate glasses can all be described using just eleven of the possible sequences.

The wide applicability of these findings in both living and non-living systems suggests that a common explanation can be applied to them all. It is now believed that the selectivity is engendered solely by the Coulombic attraction between the cations and anionic sites. Cation 1 will be bound in preference to cation 2 if the free energy of moving from the bulk water phase into the site is *more negative* for cation 1 than cation 2.

The selectivity can now be written as:

$$\Delta G_{\text{site}}(1) - \Delta G_{\text{hydration}}(1) < \Delta G_{\text{site}}(2) - \Delta G_{\text{hydration}}(2)$$

which can be arranged to give

$$\Delta G_{\text{site}}(1) - \Delta G_{\text{site}}(2) < \Delta G_{\text{hydration}}(1) - \Delta G_{\text{hydration}}(2)$$

The chief factor controlling the specificity is the field strength of the anionic site. If this is large (that is, if the site is small) then, not only will cation-site forces be large, but also the difference between them will be large. The hydration energy term may therefore be neglected and the condition for selectivity be set at:

$$\Delta G_{\text{site}}(1) - \Delta G_{\text{site}}(2) < 0.$$

Hence, the smallest cation, lithium, which can have its centre of charge nearest the site, will be bound most strongly. In previous, less quantitative, theories this was expressed by saying that the ion must lose its hydration sphere before entering a small site, so that lithium, with the smallest free ion radius, gave the best fit.

If the site is large, then the hydrated ions will all fit so that caesium, with the smallest hydrated radius, will fit best. This is more appropriately expressed by saying that in a large site the electric field strength is low so that the site–ion forces can be neglected. Thus the condition for selectivity may now be set at

$$\Delta G_{\text{hydration}}(1) - \Delta G_{\text{hydration}}(2) > 0.$$

Using values quoted in Table 2 (p. 23) it can be seen that this gives the order of fit for hydrated ions as $Cs^+ > Rb^+ > K^+ > Na^+ > Li^+$: the so-called lyotropic series.

The great advantage of using this Coulomb-attraction theory, rather than the previous ion-size fit explanation, is that it correctly predicts the intermediate selectivity patterns and that calculations are simple, using ion-site forces estimated by Coulomb's Law and measured hydration energies. The ion-fit hypothesis is incapable of producing such quantitative results. Moreover, 'hydrated-ion radii' are only approximate values at best since the experimentally derived results depend not only on the method of measurement, but also on which counter-ion is present. One final point in favour of the Coulomb-attraction theory is that the selectivity pattern of any site can be anticipated if pK_a (the acid dissociation constant) is known. If pK_a is high, meaning that the site does not easily lose a proton, then clearly the electric field strength is large and the trend will be away from the lyotropic series, whilst if pK_a is low, then the converse will hold.

Passive permeability

If the active transport system is halted by a metabolic inhibition or by cooling, then by means of tracer studies it is possible to show that the cell has a rather low passive permeability to cations, compared with the permeability to some neutral molecules. It is believed that the ions penetrate the cell by means of water-filled pores in the membranes. In general, the passive permeability of potassium ions is about ten times greater than sodium ions. In the past this has been attributed to a 'sieving' effect in the pore, with the smaller hydrated potassium ions passing through more readily. However, as current estimates of the 'equivalent pore radius' in nerve or erythrocyte are about 0·5 nm, this seems unlikely. It is now thought that much of the selectivity arises from the presence of positive charges in the pores. This may be created by the presence of calcium ions or protonated amino groups. In agreement with the hypothesis it has been shown that, in some cases, the permeability can be increased by increasing the pH, thus deprotonating the amino groups, or by adding oxalate to remove calcium ions.

Diffusion potentials and the nerve impulse

In many cells, particularly those of animals, the high intracellular potassium level is actively maintained by the 'sodium pump'. Nevertheless, the membrane is still permeable and passive ion fluxes occur under the influence of the osmotic gradients, as demonstrated by tracer experiments. Whilst potassium ions 'leak' out, sodium ions also 'leak' in, but as the mobility of the hydrated potassium is greater and the membrane is more permeable to potassium ions, the overall result is a net efflux of cations, leaving an excess of indiffusable, chiefly proteinate, anions inside the cell. Consequently, the interior of the cell is negatively charged with respect to the exterior. This electrochemical gradient then helps to reduce the outflow of cations, thus assisting the ion-pump to maintain the dynamic equilibrium.

The maintenance of a potential difference depends on continued escape of potassium ions and consequently is known as the potassium diffusion potential. The magnitude of this potential, typically about -80 mV, may readily be calculated from the Nernst equation:

$$E = 1000\frac{RT}{F} \ln \frac{[K^+]_{\text{external}}}{[K^+]_{\text{internal}}}.$$

In agreement with the prediction it has been found that if the concentration of potassium ions in the bathing fluid $[K^+]_{\text{external}}$ is increased, the transmembrane potential falls proportionately.

Nerve cells are somewhat special in that they are able rapidly and reversibly to change the permeability and thus change their polarity from the potassium diffusion, or resting potential, to a sodium diffusion or action potential. Fig.11 shows how this results in the propagation of the nerve impulse.

The increase in permeability to sodium ions results in a sodium diffusion potential which can also be calculated from the Nernst

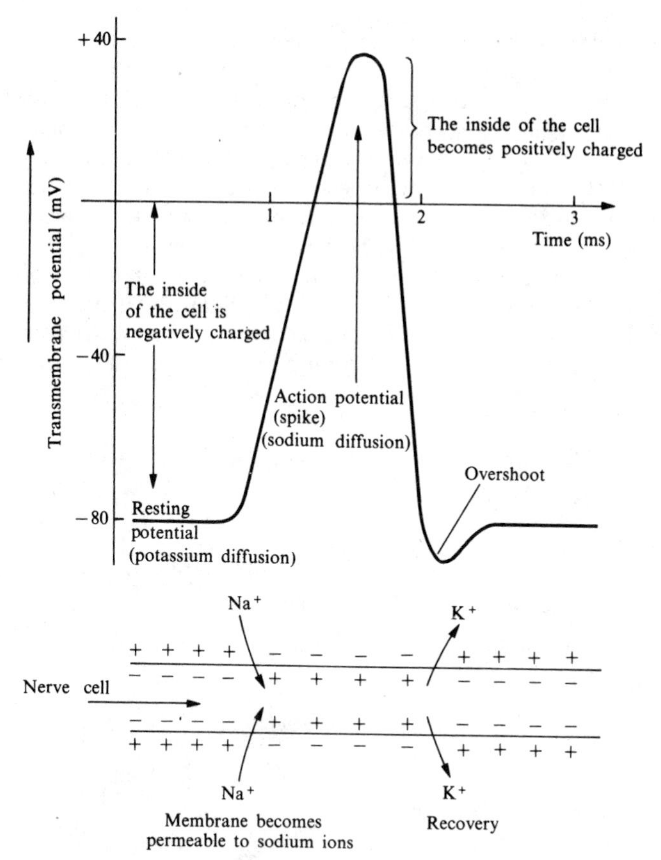

FIG.11. Diffusion potentials and the electrochemical origin of the nerve impulse.

equation. For example, in the giant axon of the squid the sodium ion concentration in the axoplasm is normally about one-tenth of the extracellular level, giving rise to a potential of about +60 mV. Again, reduction of the external sodium ion concentration results in a decrease in the sodium diffusion potential and consequently in the action potential. This could account for the poor reflexes of people suffering from hyponatraemia.

The change in permeability gives rise to a change of polarity which is the basis of the 'action potential'. In theory, this should be about 140 mV but, in practice, it never reaches this value since, as sodium ions diffuse into the cell, potassium ions begin to diffuse out, assisted, of course, by the potential difference. The result is that the net observable action potential does not normally exceed about 120 mV.

It must be emphasized that it is the *relative permeability* of the cell membrane that gives rise to the potential and the overall movement of ions is very small indeed. Indeed, the total influx of sodium ions necessary to produce an impulse in the squid axon has been estimated at 3×10^{-12} mol cm^{-2} of cell surface and since 1 cubic centimetre contains at least 1×10^{-5} mol it can be seen that a large number of impulses can be transmitted without noticeable change in the ionic concentration. Nevertheless, if the ion pump is prevented from working then the axon eventually ceases to transmit. Similarly, if the sodium ions in the bathing medium are replaced by lithium ions, the axon ceases to function since the ion-pump cannot remove lithium from the cell.

3. The alkaline earths

The chemistry of the alkaline earths

AS Table 4 shows, each alkaline earth has one more electron than its alkali-metal counterpart, so that in every case the neutral atom now has two electrons outside the closed-shell core. Consequently, the aqueous chemistry of this group is principally that of their dipositive ions, as of course was shown in Fig. 2 (p. 4). Although these ions are isoelectronic with the corresponding alkali-metal cations there are significant differences between the two groups, and it is also noticeable that the range of properties within group IIa is substantially larger. Nevertheless, these variations may still be explained largely on the basis of changes in ionic size (and hence to charge/radius ratio) on going down the group.

As expected, the ionization energies decrease as atomic size increases, mainly because the Coulombic attraction between electron and nucleus decreases with distance. Therefore beryllium, the smallest member of the group, requires the largest amount of energy to remove its two outer s-electrons. Indeed, the sum of these first two ionization energies is so large that beryllium does not always lose its valence electrons and form

TABLE 4

Some properties of the alkaline earths

		Be	Mg	Ca	Sr	Ba
Electronic configuration		(He) $2s^2$	(Ne) $3s^2$	(Ar) $4s^2$	(Kr) $5s^2$	(Xe) $6s^2$
Ionization potentials ($kJ\ mol^{-1}$)	1st	895	734	587	541	500
	2nd	1746	1443	1140	1059	959
	3rd	14000	9700	4930	4170	3580
Enthalpy of hydration of M^{2+} ion ($kJ\ mol^{-1}$)		-2494	-1924	-1580	-1485	-1276
$E^{\ominus}(M^{2+}/M)$ V		-1·85	-2·37	-2·87	-2·87	-2·90
Ionic radius (nm)		0·038	0·066	0·099	0·115	0·135
Ionic mobility† ($m^2\ s^{-1}\ V^{-1} \times 10^8$)		4·8	5·5	6·2	6·2	6·6

† See footnote to Table 2.

a simple Be^{2+} ion, but instead often forms compounds in which it has *covalent* bonds. Alternatively, it can be pictured that the ion Be^{2+} is formed but, by virtue of its extremely high charge/radius ratio (5·4 for Be^{2+} compared with 1·33 for Li^{+}), it polarizes the valence electrons of other atoms or molecules, thus creating covalent bonds.

In effect the two descriptions are equivalent, though the second describes most closely what happens when beryllium salts dissolve in water. Here, beryllium differs from the remainder of the group in that such solutions do not contain a simple aquo-ion, but instead species such as $[Be^{II}(OH_2)_3(OH^-)]^+$ predominate and the aquo-complex $[Be^{II}(OH_2)_4]^{2+}$ is stable only in strong acid solution. The loss of the proton which occurs in the absence of added acid clearly indicates the considerable weakening of the oxygen – hydrogen bond under the influence of the highly polarizing beryllium cation. As the charge/radius ratio decreases, this effect rapidly diminishes so that it is much less apparent for magnesium and can be ignored for the remainder of the group.

Like beryllium, magnesium forms a wide range of covalent compounds, of which the Grignard reagents are certainly the best known. However, these are only stable under anhydrous conditions and so have little biological or biochemical significance, whilst for the remainder of the group such compounds may safely be neglected. Despite these comments it does remain true that all the elements of Group IIa can form simple salts. Many of these are highly insoluble since the increase in ionic charge leads to a substantial increase in lattice energy which is only partly offset by the larger hydration energy of the dipositive ions. Consequently, solubility tends to decrease as the hydration energy diminishes and it is noticeable that calcium has many more insoluble or sparingly soluble salts than magnesium.

Once in solution, the alkaline earths, with the noted exception of beryllium, all form aquo-complexes. It has already become apparent that the polarizing effect of these metal ions is considerably greater than for the alkali-metal cations so that the solvent spheres associated with the alkaline-earth cations are much larger. Not only does this result in much smaller ionic mobilities for the Group IIa ions (compare the data in Tables 2, p. 23, and 4, but it also leads to greater differentiation between the individual members of this group. This is reflected in the crystalline habit of various salts. Whereas salts of beryllium and magnesium mostly crystallize as hydrates, calcium salts are less inclined to do so and the larger ions generally form anhydrous salts.

From a wide range of experimental observation it has become apparent that even in the most dilute solutions of dipositive ions, ion pairing is important. Magnesium and sulphate ions, for example, are over a hundred times more likely to exist in the form of an ion pair than are sodium and sulphate ions in equimolar solutions of the two sulphate salts. This means that cation–anion interactions will be of much greater significance in the biochemistry of magnesium and calcium than was the case of sodium and potassium and, not surprisingly, complex-formation proves to be of great importance in the biochemistry of the alkaline earths.

The cations themselves are all characteristically 'hard' Lewis acids and bind best to anionic oxygen-donor ligands such as the polyphosphates or carboxylates. Considering the effect of charge and ionic radius it is easy to predict that such ligands should form stronger complexes with magnesium than with calcium ions, though this simple pattern may easily be modified by the stereochemistry of the ligand. In general, polydentate ligands form more stable complexes with larger ions, since steric hinderance (the difficulty of packing a large number of groups round the metal ion) decreases with increasing size of the metal ion. This is illustrated by the relative stabilities of the magnesium and calcium complexes of some aminocarboxylate ligands. Thus, with the bidentate glycinate ligand, the magnesium complex is more stable but with the polydentate nitrilotriacetate and ethylenediaminetetraacetate the calcium complexes are more stable.

Methods for studying the alkaline earths *in vivo*

Since the alkaline earths form both insoluble salts and stable complexes they are more amenable to chemical analysis than are the alkali metals. The concentration of any of the alkaline-earth cations is now routinely determined by atomic absorption spectrophotometry, which has superseded the earlier technique of flame photometry since it has the advantage that it is less subject to the interference from the sodium and potassium ions found in all biological samples. For example, using atomic absorption, calcium can be accurately determined at concentrations of only 10^{-6} mol dm^{-3} (0·04 p.p.m. Ca^{2+}) in the presence of either sodium or potassium at 10^{-1} mol dm^{-3}.

It is also possible to determine metal-ion concentrations by the use of radioactive isotopes, though these have more application in the study of ion movement in metabolism. Naturally occurring beryllium consists of a single isotope, ^{9}Be, but ^{7}Be (electron capture; half-life 55 d) may be artificially prepared, though it has been little used. The study of magnesium biochemistry has been greatly hindered by the lack of a suitable long-lived isotope though ^{28}Mg (a β^- emitter, half-life 21 h), has been widely used for short-term studies. No such difficulties have been experienced for calcium, and ^{45}Ca (a β^-emitter; half-life 165 d) has been used very extensively. Naturally occurring strontium consists of a mixture of several stable isotopes but ^{90}Sr

(a β^- emitter; half-life 27·7 years) produced by uranium fission is especially important, not so much as a tracer but as a potentially lethal fall-out product. Neither barium nor radium (all of whose isotopes are active) have found much application in tracer work.

A brief search of the periodic table reveals that there is an interesting choice of elements whose dipositive ions would perhaps make useful probes by replacement of either magnesium or calcium ions. In general it is not possible to substitute one alkaline earth for another, though strontium, as ^{90}Sr, has sometimes been used as a probe for calcium. However, the use of manganese in this fashion (Mn^{2+} : radius 0·090 nm) is well established. This ion has conveniently measurable electronic absorption spectrum and magnetic properties which make it suitable for this purpose. Judged purely on the basis of size it would seem to be a better replacement for calcium than magnesium but in practice the reverse is true since like magnesium it more readily binds nitrogenous ligands. Continuing with this argument, it would appear that nickel should be even more effective, but this is generally not so as it also binds most other ligands much more strongly than does magnesium. However, it has occasionally been used most effectively, as with studies on the enzyme phosphoglucomutase. More recently considerable attention has been focused on the use of rare-earth ions, Eu^{2+} in particular, as probes for calcium. The europium ion makes an excellent substitute since, like calcium, its affinity for nitrogen donor ligands is low and the bonding in its complexes is predominantly ionic. It has the additional advantage that besides being amenable to spectroscopic and magnetic measurements, two of its isotopes, ^{151}Eu and ^{153}Eu, are Mössbauer nuclides.

The terrestrial distribution of the alkaline earths.

The alkaline earths all belong to the group of elements classed as lithophiles and as such are found concentrated in the upper regions of the earth's crust. The lightest member, beryllium, has anomalously low abundance (Fig.4; p. 12) with typical concentrations in rocks and soil being less than 10 p.p.m. By virtue of the similarity in their charge/radius ratios (Be^{2+} = 5·4, Al^{3+} = 5·3) it is usually found in conjunction with aluminium and is richest in the beryl minerals, which are mixed beryllium aluminium silicates containing up to 100 p.p.m. beryllium.

The next member of the group, magnesium, is of very much higher abundance, being mostly concentrated in the igneous rocks with a concentration there of the order of 20 000 p.p.m. Calcium is also abundant in igneous rocks, and levels of up to 50 000 p.p.m. are not uncommon. Both are found in sedimentary rocks often as relatively pure salts, for example, dolomite ($MgCa(CO_3)_2$) or gypsum ($CaSO_4.2H_2O$). The detailed distribution of these two elements represents a well documented example of bio-geochemical cycling and, moreover, one which may be understood by the application of relatively simple physico-chemical principles.

The cycling of the alkaline earths is inextricably linked with the carbon cycle since the overall concentration of dissolved metal ions depends intimately on the amount of carbon dioxide in solution. For calcium, in particular, both the processes of solution and precipitation depend, either directly or indirectly, on the fact that the hydrogen carbonate $Ca(HCO_3)_2$ is more soluble in water than the carbonate. Leaching from the primary rocks depends either on the direct dissolution of metal salts or, more important, on the extraction of the alkaline-earth cations from the silicate lattice. This process is considerably increased by the formation of the more soluble bicarbonates (Fig.12) and, since the amount of bicarbonate anion depends directly on the dissolved amount of carbon dioxide, calcium will be brought into solution or re-precipitated in response to changes in the carbon-dioxide level. One of the simplest examples, illustrated in Fig.12, occurs when underground water containing an excess of dissolved carbon dioxide reaches the surface. Here loss of carbon dioxide results in the pre-

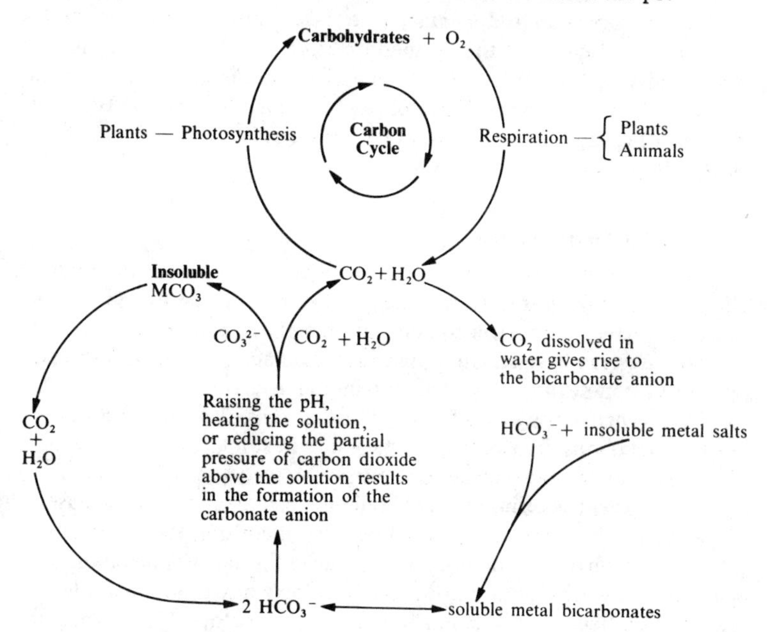

FIG.12. The bio-geochemistry of calcium and the carbon cycle.

cipitation of the calcium salt. A biotic modification of this equilibrium occurs in fresh-water lakes and ponds where the assimilation of carbon dioxide by the aquatic organisms also results in the precipitation of calcium carbonate. Another interesting example of this equilibrium shift occurs when hard water is boiled. The temporary hardness is then precipitated as 'fur' or scale as the bicarbonate is converted to carbonate.

The bulk of the dissolved salts eventually reach the sea where there is a further process of fractionation and cycling, still explicable on the basis of relative solubility. The more soluble magnesium salts remain in solution and the metal ion concentration may rise to over 1000 p.p.m. (average value 1350 p.p.m.), being limited of course by the solubility of salts such as the carbonate and phosphate. On the other hand, the calcium level rarely exceeds 400 p.p.m., though occasionally surface layers of tropical seas may have higher concentrations, indicating that they are in fact supersaturated with respect to calcium carbonate. This is important since it clearly promotes the formation of calcium-carbonate shells of certain plankton. The 'fixing' of calcium in this fashion has resulted in the spectacular formation of the immense coral reefs and islands and during the course of geological time these are converted to limestone and hence even to marble. In contrast, deposits of purely magnesian sedimentary material are rare, owing simply to the greater solubility of the magnesium salts. Similar processes may also be involved to explain the cycling of strontium and barium, though by way of contrast the presence of radium depends rather on the factors affecting the distribution of thorium, its parent in the radioactive decay scheme.

The alkaline earths *in vivo*

Of the six elements of this group only two, magnesium and calcium, have any significant biological role, all the others being toxic to various degrees. Beryllium toxicity has long been known and even sub-lethal doses can have serious consequences, since contact with the dust produces non-healing ulcers and carcinoma. It is thought to act by its ability to form strong covalent-type bonds, thus displacing other divalent metal ions, particularly magnesium. In support of this view it has been shown that even at concentrations as low as 10^{-6} mol dm^{-3} it inhibits the magnesium-potentiated enzyme, alkaline phosphatase. Low levels of strontium are found in most organisms, though the inclusion is probably accidental, occurring via the calcium metabolic pathway, and strontium is not thought to play any functional role. It is not particularly toxic, but the radioactive isotope ^{90}Sr is dangerous since it accumulates in the bone, where it causes considerable damage to the marrow and eventually produces leukaemia. Whilst barium is slightly more abundant than strontium it too has no biological significance (as far as is known) except in so far as its soluble salts are extremely toxic, owing to interference with calcium function. In contrast the extremely insoluble sulphate salt is completely harmless and is ingested for diagnostic purposes as it is opaque to X-rays,

It now appears that all organisms have an absolute requirement for magnesium. In plants magnesium deficiency is characterized by chlorosis, although the converse is not necessarily true.† Because of the relative solubility of magnesium salts they are easily leached from the soil which may thus become magnesium deficient and require regular supplementation. On the other hand, on the so-called serpentine soils the flora must be tolerant to the high levels of magnesium and other ions found in the soil-waters. Animals, particularly herbivores, usually acquire sufficient magnesium in their diets to make an inorganic supplement unnecessary, and magnesium deficiency is rare. It is most commonly found as a side-effect of a general dietary imbalance, as in the protein-deficiency condition 'Kwashiorkor', or in cases of chronic alcoholism. It is relatively easy to produce magnesium deficiency artificially and the symptoms include excessive nervous irritability and even tetany (intermittent muscular spasms). In contrast, an excess of magnesium acts as a purgative and can produce muscular relaxation.

A typical adult human being requires 200–300 mg magnesium daily, in order to maintain the body pool at the normal level of approximately 20 g. About half the total magnesium ingested is incorporated into the bone, where it is believed to replace calcium at occasional sites, thus modifying and strengthening the bone structure. Although magnesium phosphate is slightly more soluble than the corresponding calcium salt, it is still 'insoluble' for most purposes and, in the presence of excess phosphate, magnesium may be precipitated. This most commonly occurs in infections of the urinary tract, where the double salt $NH_4Mg(PO_4)6H_2O$, 'the coffin plate crystal' can cause serious obstruction. Almost all of the remainder of the bodily magnesium is located intracellularly (see Table 5).

Plants can be broadly classified as calcifuges (calcium-hating), which can thrive only in areas where the soil has a low calcium content, or calcicoles, which flourish on soils rich in calcium. The basis of the distinction is complex, and is a result not only of calcium levels, but also of the related pH effects. The use of lime (calcium hydroxide) as a fertilizer is widespread and historical. Again much of the beneficial effect comes not from the increase in calcium concentration but rather

† Chlorosis is a general term describing an abnormal yellowing of the leaves due to a deficiency of chlorophyll.

TABLE 5

The concentrations (p.p.m.) of magnesium and calcium in rocks, water, and living materials

	Mg^{2+}	Ca^{2+}	Ratio †
Igneous rocks‡	25 000	40 000	1 : 1
Sedimentary rocks‡	10 000	30 000	1 : 1·8
Limestone	3000	300 000	1 : 182
Sea-water	1400	400	1 : 0·2
Salmon (whole blood)	24	240	1 : 6·0
Myxine (whole blood)	440	240	1 : 0·28
Man (whole blood)	36	60	1 : 1
(plasma)	36	120	1 : 2
(interstitial fluid)	50	120	1 : 1·44
(cell fluid)	2640	4	1 : 0·0009
(muscle) §	75	1400	1 : 0·09
(nerve) §	750	280	1 : 0·23

† Calculated as relative *numbers* of atoms (corrected for relative atomic masses).
‡ Average values expressed as p.p.m. of dry weight.
§ p.p.m. of whole tissue.

from the increase in the alkalinity of the soil. Unfortunately, excessive liming has the effect of precipitating many of the trace elements and preventing them from being taken up and so is actually harmful. The bulk of calcium in plants is found as calcium pectate† in the cell walls. Calcium oxalate is also found in many plants and seems to serve the dual purpose of removing unwanted organic acid and at the same time reducing the free calcium level in the plant.

In man and other animals the overall control of the calcium level is achieved by balancing intake and excretion. Despite the fact that calcium is readily available in most diets, deficiency diseases are not un-

† Pectic acid is a polymer of galacturonic acid.

common, as there is some difficulty in the actual absorption of this element. One major difficulty is the tendency for the calcium to be precipitated by many anions present in food. The main interference in this fashion comes from the phosphate ion so that calcium and phosphate balances are inseparable. Only calcium dihydrogenphosphate, $Ca(H_2PO_4)_2$, is sufficiently soluble to maintain the level necessary for efficient absorption. Unfortunately this salt is stable only under the acid conditions found in the stomach while the actual process of absorption occurs in the duodenum. Therefore absorption is restricted to the relatively short time before the acid chyme discharged by the stomach is neutralized by the alkaline bile. Clearly, high-protein diets which are phosphate-rich, will be unfavourable to calcium absorption, as will be the use of gastric antacids which raise the pH and decrease the concentration of $H_2PO_4^-$. The actual process of absorption is by active transport of calcium. The efficiency of this process is controlled by Vitamin D and the parathyroid hormones, which also control the formation and resorption of the bones.

Hypercalcaemia is widely recognized, and the symptoms include not only the formation of urinary calculi but also a general thickening of the bones and calcification of the cartilages, whilst excessive levels of calcium in the plasma eventually cause the heart to stop in systole (contraction). If the overall calcium level drops then oesteoclasts bring about the resorption of the bone and a mobilization of calcium, in a process mediated again by Vitamin D and parathormone.† In cases of prolonged dietary deficiency this results in a variety of abnormalities in the bones. Low levels of plasma calcium result in excessive irritability of nerves and muscles and may produce tetany.

Despite the fact that the overall distribution of both magnesium and calcium is controlled by a specific active transport system which accumulates magnesium intercellularly and leaves calcium ions outside the cell, the roles of these two metals are quite unlike those of the corresponding alkali metals whose electrochemical and osmotic functions depend on the maintenance of such gradients. Instead the alkaline-earth cations function by the formation of a range of insoluble salts and stable complexes. The structural role of calcium ions has already been touched on and this function arises naturally from the insolubility of its phosphate and carbonate salts. However, it is perhaps less well known that calcium also acts as a neuromuscular trigger by a process related to the formation of phosphate deposits, namely, by acting as a bridge between the phosphate groups attached to the protein strands in striated muscle.

† Osteoclasts are large multi-nucleated cells functioning to absorb and remove osseous tissue.

Magnesium, on the other hand, is best known as its complex, chlorophyll, which is the primary photosynthetic pigment of all green plants (p. 55). It also has an important role in enzyme activitation.

Magnesium and enzymes

The role of magnesium, or indeed of any metal ion, as an enzyme activator can be discussed under the four headings: structure-promoter, substrate binder, Lewis acid, and electron carrier. Since it is well known that in aqueous solution magnesium is restricted to only one oxidation state, namely, +2, it can quite clearly be excluded from the role of an electron carrier in a redox process. In fact, as is discussed later, this role is reserved for the transition-metal ions. The role of structure-promoter is also a relatively uncommon one for magnesium, though a few example are known. The enzyme pyruvate decarboxylase, which requires thiamin pyrophosphate as a coenzyme, exhibits activity only in the presence of a dipositive metal ion, generally magnesium. This ion may be replaced by several other metal ions, notably the Mn^{2+}, but the activity of the resulting apoenzyme–metal–thiamin pyrophosphate complex is independent of the nature of the metal. This may be taken to imply that the metal ion acts merely as a structure-promoter, in this case as a bridge, ensuring the structural integrity of the holo-enzyme.† There appear to be few, if any, magnesium-potentiated enzymes in which the metal ion acts solely as a substrate binder, linking the enzyme and its substrate

By far the largest group of enzymes which require magnesium in order to exhibit their full activity are those which catalyse the hydrolysi or cleavage of various polyphosphates. This class includes enzymes such as alkaline phosphatase, the ATPase of the 'sodium pump'; hexokinase; and deoxyribonuclease (DNase I); yet despite their diversity it seems likely that in every case the role of the magnesium ion is the same. It is now believed that the metal acts by binding to the phosphate residues, where it functions as a Lewis acid, polarizing these groups and hence increasing the possibility of nucleophilic attack on the terminal phosphorous atom, as is shown in Fig. 13.

This mechanism is supported by studies on the metal-ion catalysed hydrolysis of pyrophosphates. In this reaction the nucleophile is a water molecule, and by using labelled [^{18}O] water, it is possible to demonstrate that there is a direct attack by the nucleophile on the terminal phosphorus atom, because the liberated phosphate ion contains ^{18}O which can be incorporated only at this stage, since the exchange of

† Enzymes are proteins but may also require cofactors or coenzymes before catalytic activity can be observed. The intact protein–cofactor–coenzyme unit is known as the holo-enzyme and the protein minus these units is known as the apo-enzyme.

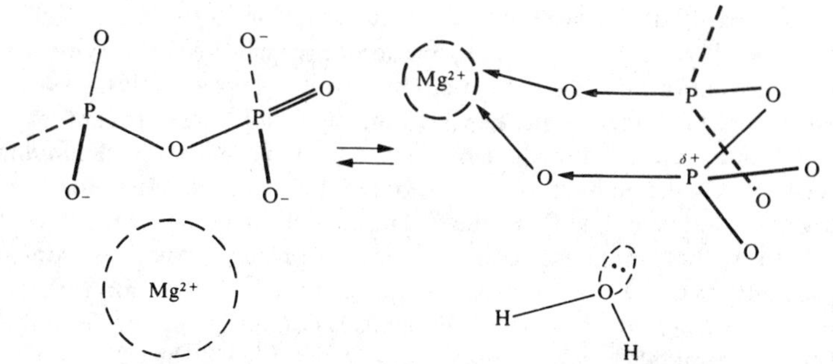

FIG.13. A possible role for magnesium in the hydrolysis of polyphosphates. The formation of the magnesium-polyphosphate complex could increase the rate of hydrolysis by modifying the shape, reducing the steric hindrance, and by polarizing the molecule, thus increasing the likelihood of nucleophilic attack by the water on the terminal phosphorus.

oxygen between the phosphate ion and water is slow. It seems likely that a similar single-step mechanism operates in the hexokinase catalysed reaction, though in this case the nucleophile is no longer a water molecule, but the hydroxyl group attached to the 6-position of the hexose ring.

In contrast, the alkaline-phosphatase catalysed hydrolysis of ATP occurs in two distinct steps. The first stage of the reaction employs the hydroxyl group of an adjacent serine residue as the nucleophile, which results in the formation of a monophosphate ester. This is then rapidly hydrolysed by water in the second stage of the reaction. Many similar examples of the two-stage hydrolysis of ATP are known; for example the ATPase catalysed hydrolysis of ATP also occurs sequentially. The first stage of the reaction employs the carboxylate group of an amino-acid side-chain as the nucleophile, which results in the formation of an acyl phosphate (mixed anhydride) intermediate, which is itself then hydrolysed. Nevertheless, the unifying factor in all these apparently diverse reactions is the formation of the magnesium–polyphosphate complex.

The antagonism between magnesium and calcium ions

In contrast with the many magnesium-activated enzymes, examples of calcium-requiring ones are relatively few. Moreover, in those enzymes which do require calcium it seems that the metal plays only a structure-promoting role. A typical example is the endo-glucosidase, α-amylase, which catalyses the hydrolysis of α-1,4 glucosidic linkages. In the absence of calcium ions the enzyme is completely inactive and recent experiments suggest that the principal function of the metal is to maintain the structural integrity of the protein chain. One exception to this general rule is the role of calcium ions in the formation of trypsin from the zymogen trypsinogen. In this case the metal ion actually plays a catalytic role, accelerating cleavage at one peptide bond and retarding it at another.

Of more interest though, is the fact that the presence of calcium ions actually inhibits many magnesium-potentiated enzymes. As a first step towards understanding this antagonism, it would be convenient to be able to predict whether calcium ions could displace magnesium from any particular site, but unhappily the simple Coulombic-attraction theory which works so well for the alkali metals is not easily applicable to the Group IIA metal ions. The greater polarizing effect of the +2 ions makes non-Coulomb forces, particularly those of the ion induced-dipole type, important, especially for sites of low field strength. Of course, steric influences may also be considerable. As a result it is usually a matter for experimental determination as to whether calcium ions will displace magnesium ions from an enzyme. However, given that calcium may displace magnesium, or at least bind equally strongly, it is still necessary to explain why the calcium ion is catalytically inactive.

This question has been examined in considerable detail, particularly in the respect to which the two ions can bring about the hydrolysis of ATP. The rate of this reaction is markedly increased by the presence of magnesium ions yet calcium ions have little catalytic effect and actually inhibit the increase in rate caused by the magnesium. There appears to be combination of several factors involved in this discrimination. First, calcium ions are larger and hence less polarizing than magnesium or, in other words, they are weaker Lewis acids. Second, the rate of exchange of ligands bound to calcium is generally greater (by a factor of at least 100) than those bound to magnesium. As a consequence, the ligand spends less time bound to the calcium ion than to the magnesium, so the time available for catalytic reaction is reduced proportionately. Third, the stereochemical requirements of the two ions may be dissimilar and this will affect the relative position of reacting groups and hence the course of the reaction.

Magnesium and chlorophyll

The best-known functional compound of magnesium is chlorophyll (Fig.14). The role of magnesium in chlorophyll is not immediately obvious, but apart from acting as a link between layers of chlorophyll aggregates it certainly has an effect on the properties of the porphyrin ligand and by acting as Lewis acid will modify the electron distribution in the molecule and thus alter both its spectroscopic properties (that is, the wavelength at which it absorbs the maximum amount of light) and also its electron-transporting powers.

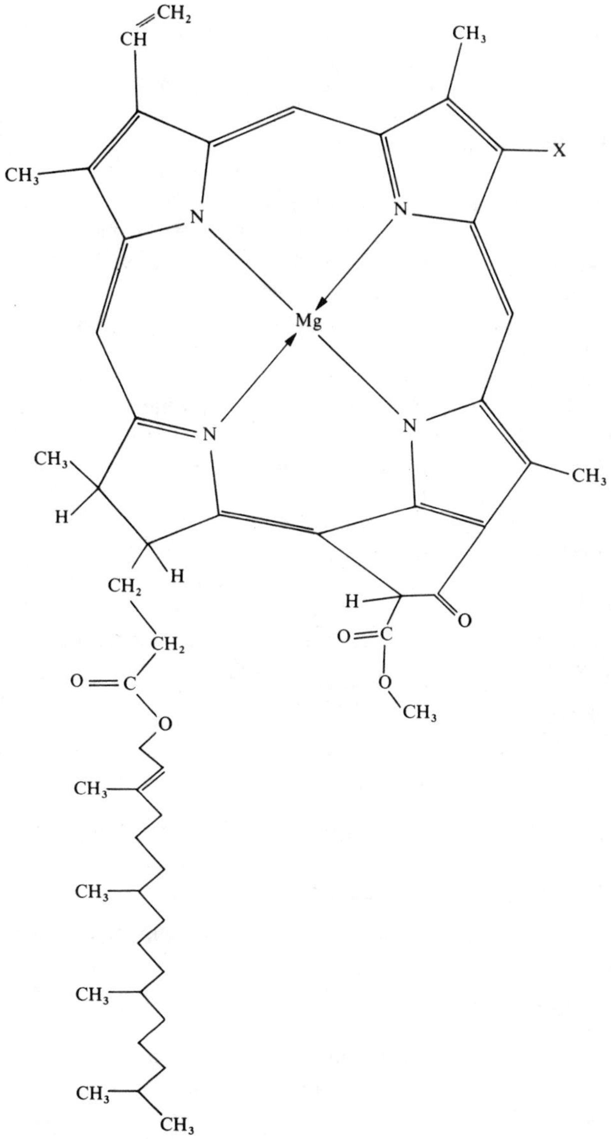

FIG.14. The structure of chlorophyll.

Extracellular calcium

As the data in Table 5 show, the vast majority of calcium in man is extracellular and though most of this is in turn localized in the skeleton there still exists a small but important calcium pool in the plasma. Obviously the total concentration of calcium in this fluid should be limited by the solubility of its least soluble salts, in this case the hydrogen phosphate and the carbonate, yet experiments clearly show that the calcium level in solution exceeds these maximum permitted levels. The explanation is fairly straightforward. Much of the plasma calcium is complexed, particularly by the plasma proteins, and hence is held in solution despite the level of the precipitating anions.

The plasma calcium represents a mobile fraction which provides the link between absorption and deposition in the bone. In addition plasma calcium performs certain other functions, the most important of which is its role in the clotting of blood. Blood-clots are basically formed by an insoluble matrix of the protein, fibrin, which in turn is formed from a more soluble precursor fibrinogen by the action of thrombin. The process is complex and thrombin itself is derived from other precursors and prothrombin. Calcium is required at several stages though its function is still not entirely clear.

In man the primary control of calcium levels is achieved by limiting absorption under the influence of the parathyroid hormones and Vitamin D. The absorbed calcium is then translocated via the plasma proteins, with the hormones and vitamin implicated in the longer-term control of calcium balance through their influence on the formation and resorption of bone, which in man represents the major portion ($>$95 per cent) of all body calcium and which acts as the principal calcium reserve or depository of excess calcium.

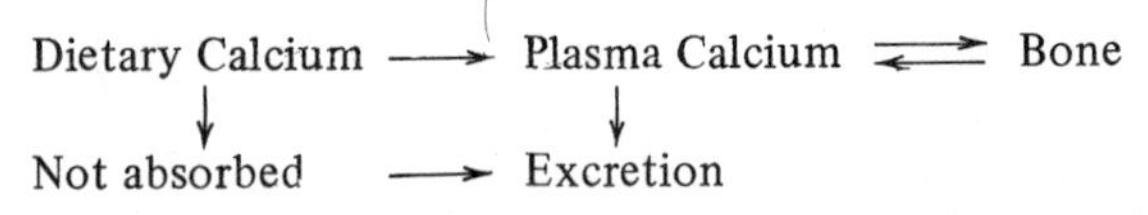

The formation and resorption of bone is controlled by specific cells. It is thought that the bone is built up from a precipitate of octacalcium phosphate $Ca_8H_2(PO_4)_6 6H_2O$ laid down on a framework of the protein collagen by cells known as osteoblasts. The salt is then slowly converted to the normal form of bone, calcium hydroxyapatite. It is notable that bone does not normally contain the tribasic phosphate $Ca_3(PO_4)_2$ as this is more soluble than the hydroxyapatite, though in cases of hypercalcaemia deposits of $Ca_3(PO_4)_2$ are formed as stones in the bladder and kidney.

Most other organisms seem to require calcium on account of its general structure-promoting role, though its function ranges from acting as a bridge between organic components of the cell wall to the formation of a skeleton proper. A particularly interesting example of the former role is the phenomenon of 'surface precipitation' exhibited by *Arbacia* eggs. In normal sea-water if the cell membrane of such an egg is injured, then a new membrane is quickly formed. However, in artificial calcium-free saline this process does not occur. Of course the formation of calcium-carbonate shells is also a widespread phenomenon particularly amongst marine organisms, though it is less common in fresh-water species, owing to the lower levels of calcium salts in this environment.

Intracellular calcium

The data shown in Table 5 refer only to the total analytical concentrations for cellular calcium. It is now believed that the concentration of free calcium ions within the cell may be as low as 10^{-8} mol dm^{-3} (40 p.p.b) and that the majority of the calcium is bound to the cellular membranes. The low intracellular level of free calcium is a prerequisite of metabolic efficiency, as calcium ions inhibit most of the magnesium-activated enzymes, which are located principally inside the cell. In particular it has been shown that moderate intracellular levels of calcium inhibit the sodium pump. In muscle cells the mechanism of contraction requires a sudden increase in the concentration of calcium ions; this can most easily be achieved if the initial level of free calcium is kept low. Generally the membrane provides a convenient store for the ions and these ions restrict the permeability of the membrane to other metal ions in the manner described previously.

If the Nernst equation is applied to this system it gives rise to two results, both of which are contradicted by experimental measurements. Using the observed resting potential and the measured extracellular concentration, the intracellular level is predicted to be 1000 times greater than outside. Alternatively, using the measured free ion concentrations, a calcium diffusion potential of about +150 mV is calculated. The reason for these peculiar results is that the Nernst equation is applicable only if the ions are relatively free to diffuse across the membrane. Experiments with ^{45}Ca have confirmed the obvious conclusion that calcium ions are not free to move in this fashion. For instance, erythrocytes placed in plasma containing a high concentration of ^{45}Ca and then treated with a metabolic inhibitor gained only 2 per cent of this tracer over a period of 6 days. Clearly then, membrane impermeability accounts for a large proportion of the calcium concentration gradient. Nevertheless, over the 100-day life span of the average erythrocyte, some changes in concentration would be expected to occur, and in order to account for the absence of such effects in metabolically active cells it is necessary to propose the presence of an active calcium transport system. There seems little doubt now that such a 'pump' actually exists and that it is driven by the hydrolysis of ATP brought about by a calcium-activated ATPase. This enzyme has been identified in membrane fragments of the erythrocyte and the stoichiometry of the system determined at 1 Ca^{2+} : 1 ATP. Interestingly, the system will transport strontium though it is inhibited by other Group IIA ions.

Calcium and muscular contraction

Muscle tissue can be divided into two broad categories: smooth muscle, which is generally responsible for involuntary functions; and striated muscle, the muscle of the arms, legs, etc., which is under voluntary control. The morphology of striated muscle has been intensively studied and it is generally accepted that the basic unit responsible for contractility is the sarcomere, shown in Fig.15. Whilst the detailed mechanism of contraction has not yet been fully elucidated, enough is known to be able to put forward a plausible model.

Initial stimulation occurs when acetylcholine is released from the end foot of the neurone.† This crosses the gap between the nerve and the motor end-plate and there penetrates the plasmalemma.‡ Since acetylcholine carries a positive charge it disturbs the resting potential of the membrane and as a result a wave of action potential spreads along the muscle. This results first in the release of calcium from the plasmalemma into the cell. This arises since the resting potential normally acts to separate cations and anions by a field effect and the reversal of this increases ion pairing with the result that the calcium becomes bound to a lipophilic anion and subsequently diffuses through the membrane.

This affects an interior membrane, the sarcoplasmic reticulum, which, in the relaxed state of muscle, binds the intracellular calcium. The calcium is now released and readily forms a link between the phosphate groups on actin and myosin (Fig.15). This neutralizes the negative charge of the myosin–ATP unit so that it is no longer repelled from the myosin–ATPas by electrostatic repulsion. Therefore the myosin side-chain rearranges itself to a more stable α-helix and since it is linked to the actin filament, this results in relative movement of the actomysin complex and muscular contraction. The link is then broken by hydrolysis of myosin–ATP to myosin–ADP and inorganic phosphate with consequent release of Ca^{2+}. Finally the myosin is restored to the resting state by rephosphorylation using free ATP. It can be pictured that this process occurs simultaneously at numerous points in the muscle and is repeated many times during the contraction process. Contraction ceases when the calcium level is reduced

† The nerve cell; the morphological unit of the nervous system.
‡ The plasmalemma is a membrane enclosing the plasma.

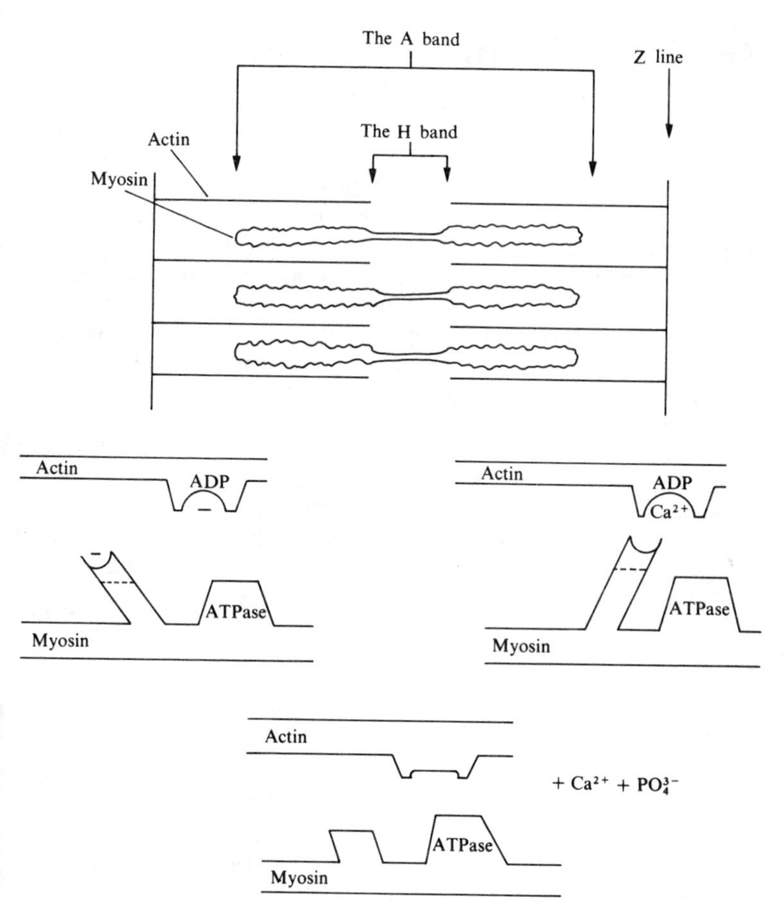

FIG.15. The sarcomere, the basic unit of striated muscle, consists of bands of overlapping protein fibres. Calcium is believed to act as a bridge between the actin and myosin strands, bringing about movement between them in a 'ratchet' fashion.

by rebinding to the sarcoplasmic reticulum after restoration of the membrane resting potential. Of course this model is much simplified and a more detailed one must account for the calcium flux into and out of the cell during excitation and relaxation.

4. Zinc, Cadmium, and Mercury

Periodic classification and chemical properties

IN most modern periodic classification schemes the three elements which comprise Group IIB are to be found at the end of the d-block transition series, just after copper, silver, and gold. Whilst they have some similarity to copper, they might equally well be compared to the alkaline earths – with which they share an ns^2 valence electron configuration – and so be placed at the other end of the periodic table. However, this latter comparison must not be over-emphasized as there are some distinct group differences.

Notwithstanding these comments, the dipositive state still dominates the aqueous chemistry of the Group IIB elements. There is no evidence that they ever form compounds in which they are tripositive, nor indeed is this likely, considering the magnitude of the third ionization energies (Table 6). However, unlike the alkaline earths, which are strictly dipositive, singly charged cations of all three members of Group IIB are

TABLE 6

Some properties of the Group IIB elements

		Zn	Cd	Hg
Electronic configuration		(Ar) $3d^{10}\,4s^2$	(Kr) $4d^{10}\,5s^2$	(Xe) $5d^{10}\,6s^2$
Ionization potentials ($kJ\ mol^{-1}$)	1st	910	862	1010
	2nd	1740	1630	1810
	3rd	3880	3680	3330
Enthalphy of hydration of M^{2+} ($kJ\ mol^{-1}$)		-2046	-1807	-1824
$E^{\ominus}(M^{2+}/M)$ V		-0·76	-0·40	+0·85
Ionic radius (nm)		0·069	0·092	0·093
Ionic mobility† ($m^2\ s^{-1}\ V^{-1} \times 10^8$)		5·6	5·47	6·56 (Hg^{2+}) 7·11 $(Hg_2)^{2+}$
Average abundance in rocks and soil (p.p.m. of dry weight)		20~100	0·5~1·0	0·5~1·0

† See footnote to Table 2.

known. Admittedly Zn^+ and Cd^+ are mainly confined to ionic melts, which can hardly be considered suitable biochemical environments, but Hg(I) is well known, even in aqueous solution. The so-called mercurous cation is not a simple monomeric Hg^+ ion, but is in fact, a unique dimeric species, $^+Hg-Hg^+$, with a strong metal–metal bond. Although mercury (I) salts are known in aqueous solution the $(Hg^+)_2$ ion is somewhat unstable towards disproportionation to mercury and the mercuric ion, as the appropriate redox potentials show. However, the equilibrium is easily shifted and by the addition of excess mercury metal the $(Hg^+)_2$ ion is re-formed spontaneously.

$$Hg^{2+} + 2e \rightarrow Hg_{(l)}, \quad E^{\ominus} = 0{\cdot}854 \text{ V}$$

$$2Hg^{2+} + 2e \rightarrow (Hg^+)_2, \quad E^{\ominus} = 0{\cdot}920 \text{ V}$$

$$(Hg^+)_2 + 2e \rightarrow 2Hg_{(l)}, \quad E^{\ominus} = 0{\cdot}789 \text{ V}$$

whence $$Hg^{2+} + Hg_{(l)} \rightarrow (Hg^+)_2, \quad E^{\ominus} = +0{\cdot}131 \text{ V}.$$

The Group IIB metals differ from their Group IIA counterparts by having a filled set of d-orbitals interposed between the core and the valence electrons. The diffuse nature of the d-electrons means that they provide only relatively poor shielding for the increased nuclear charge, so that the ionization energies of zinc, cadmium, and mercury are higher than those of calcium, strontium, and barium, respectively (Tables 4 and 6). Conversely, then, the cations Zn^{2+}, Cd^{2+}, and Hg^{2+} have such an avidity for electrons that complete ionization is unfavourable, with the result that complexes of these metals have considerable covalent character.

Electron sharing will be energetically least expensive, and hence most favourable, if the ligand involved is relatively large and polarizable. It is not surprising, therefore, to find that the Group IIB metals prefer to form compounds with ligands such as the chloride ion, or those having phosphorus or sulphur donor atoms, whilst the Group IIA ions preferentially form complexes with fluoride ion or hard, non-polarizable, oxygen-donor ligands. The difference between the two groups is elegantly illustrated by the equilibrium shown below:

$$\underset{\text{Blue}}{6H_2O + [Co^{II}Cl_4]^{2-}} \underset{Ca^{2+}}{\overset{Zn^{2+}}{\rightleftharpoons}} \underset{\text{Pink}}{[Co^{II}(OH_2)_6]^{2+} + 4Cl^-}$$

Calcium ions move the equilibrium to the left by preferentially binding water molecules, whereas zinc binds the chloride ions and moves it to the right.

Many other group differences can be explained in a similar fashion. For example, the sulphides of Group IIA are all soluble in water because these metal ions bind oxygen ligands in preference to sulphur donors, but the converse is true for Group IIB metal ions, and so their sulphides are all insoluble in water. This is the rationale for the distinguishing test with hydrogen sulphide in the classical qualitative inorganic analysis schemes. Again, the oxides of Group IIA are all much more stable than those of Group IIB, and indeed mercuric oxide is readily dissociated simply by heating.

The variations within Group IIB are more difficult to rationalize since changes in the parameters usually considered, ionization energies and redox potentials, are irregular. However, it is possible to say that, although compounds of zinc and cadmium have about the same proportion of ionic character, mercury compounds are considerably less ionic, so that mercury (II) chloride is estimated to be almost 50 per cent covalent. The increase in covalency may explain the relative stability of the oxides. Zinc and cadmium oxides are stable and are used as paint pigments but mercury(II) oxide is decomposed by heating. Finally, the trend towards increasing covalency is most vividly demonstrated by the stability of the metal alkyls in water. Zinc and cadmium compounds react vigorously and are immediately hydrolysed but the corresponding compounds of mercury are resistant to hydrolysis, a fact of considerable ecological importance.

Methods for studying the Group IIB metals *in vivo*

Since tissues normally contain fairly large amounts of zinc, this element can usually be estimated quite satisfactorily by conventional analysis. Much use has been made of dithizone (diphenylthiocarbazone) as a reagent, both for extracting zinc and its subsequent spectrophotometric estimation. A great deal of attention has also been paid to the determination of both cadmium and mercury in living material in studies on their toxicity. Whilst both metals can generally be estimated quite satisfactorily by classical analysis, mercury in particular may be accurately determined at extremely low concentrations by means of neutron activation analysis. The sample of mercury-containing material is irradiated in a reactor using thermal neutrons. This induces radiative capture reactions, and by this process ^{196}Hg is converted to ^{197}Hg. The product is unstable (half-life 65 h) and the mercury content of the sample can be measured by quantitative detection of the γ radiation produced during its decay. As little as 10^{-9} g can be detected, and 10^{-8} g can be estimated accurately. This method has the advantage that it is non-destructive, so that the sample can be used for further tests.

Much information about zinc metabolism has been derived from tracer studies using ^{65}Zn (a positron emitter; half-life 244 d), and ^{109}Cd (electron capture; half-life 470 d) is potentially useful, though it has not yet been greatly employed.

Zinc-potentiated enzymes are numerous, but unfortunately the closed-shell d^{10}-configuration of the zinc ion means that it has no electronic transitions which might be spectroscopically useful, but in several cases the detailed role of the zinc ion has been deduced by replacing it with cobalt. Despite the difference in sizes (the radii are 0·069 nm for zinc and 0·082 nm for cobalt) the dipositive cobalt ion makes an excellent probe since many of the enzymes retain some activity after the substitution, suggesting that the mechanistic deductions may be valid. Not only this, but the Co^{2+} ion has well-understood spectroscopic and magnetic properties.

The terrestrial distribution of the Group IIB metals

Although all three elements are of relatively low overall natural abundance (Table 6), they are well known since they occur in high local concentrations and are easily isolated from their ores. In keeping with their chemistry they are all found as sulphides, though zinc, with the highest ionic character of the group, is also found as the carbonate (calamine), the oxide, and the silicate, whilst mercury (though apparently not cadmium) is also found as the native metal.

The three elements are all relatively mobile and much is known about the detailed mechanism of their cycling. For zinc the process is easy to describe in terms of simple chemical steps. The primary sulphide is first oxidized to the sulphate and the concomitant increase in solubility accelerates the leaching from the rocks. The zinc in solution is subsequently re-precipitated, mainly as the carbonate, which under the influence of metamorphic events is decomposed to the oxide.

The cycle is modified by the accumulation of zinc by plants and this is reflected in zinc content of certain bioliths, for example coal ash may contain up to 10 000 p.p.m. zinc, over 100 times the average level in rocks and soil. Similar cycles can be described for the other two elements, even to the extent that metallic mercury has been noted as a residue from condensates of coal-gas production.

The Group IIb elements *in vivo*

Of this group cadmium and mercury are both highly toxic and zinc now appears to be an essential micro-nutrient for all organisms, both plant and animal. As a result of the variable distribution of zinc in soils due to leaching, zinc deficiency in plants is well known, though the converse is not true, and even on soils rich in zinc toxicity symptoms are most infrequent. Zinc deficiency is not just a function of a low level of zinc in the soil, but may also become apparent on highly organic or calcareous soils where the metal ion is immobilized by complex-formation and precipitation. The zinc status of animals depends in turn on that of plants, and as a consequence is also highly variable.

The average daily human diet contains about 20 mg zinc; studies with ^{65}Zn show that about half of this is absorbed. Little is known about the mode of absorption except that it occurs primarily in the distal portion of the small intestine. *In vitro*, this process is inhibited by cyanide ions, suggesting that a process of active transport might be involved, but dinitrophenol has no effect, so the energy source cannot be oxidative phosphorylation. Also no carrier molecule has yet been identified. Both Cd^{2+} and Cu^{2+} interfere with zinc absorption, as does a vegetarian diet containing large amounts of phytic acid,† which forms an insoluble complex with zinc ions.

The average adult human being contains about 2 g of the element, of which one-quarter to one-third is to be found in the skin and bones. Intravenously injected ^{65}Zn is rapidly translocated to these regions, and it is believed that this portion of the body pool acts as a buffer and a reserve. A prolonged dietary deficiency results in the mobilization of this fraction, producing the typical symptoms of skin lesions and skeletal abnormalities. The general concentration of tissue zinc is low, and the remaining zinc is to be found principally in the pancreas, the eye, and the male sex organs. Zinc is also found in spermatozoa, and the spermicidal effect of cadmium may be partly due to the displacement of zinc. Zinc deficiency results in other disturbances of the reproductive systems such as hypogonadism (failure of testicular development) and also in congenital defects of offspring, probably by a failure of nucleic-acid synthesis. Dwarfism is one result of gross deficiency from birth.

An interesting, but as yet unexplained, facet of zinc biochemistry is its role in healing wounds. This process is speeded up by treatment with zinc, and it has been shown that after major surgery zinc is rapidly mobilized from the skin and bones, hypozincuria sometimes resulting unless supplementary zinc is supplied. On a lesser scale, topical applications of zinc salts increase the rate of healing of bed-sores and the homely zinc and castor-oil preparation has a long and useful history as a healing ointment.

About half the body zinc is to be found in the blood, divided unequally between the plasma, 12–20 per cent, the erythrocytes, 75–80 per cent, and the leucocytes, about 3 per cent. Plasma zinc exists mainly as Zn^{2+}–protein complexes, bound mainly by the sulphydryl and imidazol group of cysteine and histidine. The erythrocyte fraction is found chiefly in the enzyme carbonic anhydrase, which will be discussed in more detail

† Phytic acid is the common name for inositol hexaphosphoric acid. Its mixed Mg^{2+}–Ca^{2+} salt finds use as a dietary supplement, providing Ca^{2+}, inositol, and phosphate.

later. The small amount of zinc in the leucocyte fraction reflects their relative scarcity since an individual leucocyte contains about 25 times more zinc than an erythrocyte.

Zinc function

The many functions of zinc can be described as either structure-promotion or enzyme activation, though the two roles are not entirely separate since in a few enzymes zinc is believed to act by stabilizing the protein structure. A similar function has been ascribed to pancreatic zinc which is largely, though not completely, confined to specific cells, the so called β-cells. These cells are responsible for storing insulin and the zinc is believed to stabilize the insulin molecule. This hypothesis is amply supported by experiments *in vitro* which show that insulin can be crystallized only if zinc or a certain few other M^{2+} ions are first added to the solution. The zinc ions do not appear to be incorporated chemically, but instead seem to be loosely absorbed on the surface of the molecule, and the more zinc there is present the longer the crystalline insulin retains its activity.

It has been suggested that *in vivo* the degranulation of the β-cell and the release of insulin, which is also accompanied by a reduction in the cellular zinc, is a result of competitive binding by histidine. Certainly, a form of artificial diabetes can be produced by injection of dithizone, which binds the zinc and interferes with the stabilization of the insulin.

$$\begin{array}{c|c} [\beta\text{-cell} - \text{Insulin} - Zn^{2+}] + \text{Histidine} & \text{Granulated} \\ \downarrow\uparrow & \uparrow\downarrow \\ [\text{Histidine} - Zn^{2+}] + \text{Insulin} + \beta\text{-cell} & \text{Degranulated} \end{array}$$

High concentrations of zinc are also found in the choroid region of the eye, up to 15 per cent in the eye of the fox, and it is believed that these zinc ions act as a bridge, binding the retina in position, since injections of dithizone into the eye produce immediate retinal detachment and blindness. Zinc also plays another role in vision being an activator for the enzyme retinene reductase.

Zinc-potentiated enzymes

The various roles of zinc ions in the many enzymes in which they are found can all be described under the three headings of structure-promoter substrate binder, and Lewis acid, as discussed previously for magnesium. (Since zinc forms only +2 ions under biological conditions, it too can be eliminated from redox processes.) Liver alcohol dehydrogenase is a particularly interesting enzyme as it has been shown to contain two functionally distinct types of Zn^{2+} ions. The active form of the enzyme is a tetramer and this contains four zinc ions, two of which are relatively easil removed whereupon the activity of the enzyme is lost, though the structu ral integrity of the tetramer is retained. It is believed that these two zinc ions lie in active sites where they function by binding the substrate. In keeping with this suggestion it has been estimated that there are two activ sites per enzyme molecule. The remaining two zinc ions are firmly bound and removal of these leads to dissociation into monomeric units. Clearly these ions must function as structure-promoters of considerable importan

The role of magnesium in the enzymic hydrolysis of ATP was discusse in Chapter 3, where it was suggested that the metal ion provided a bindin site for the terminal phosphate group, thus enhancing the possibility of nucleophilic attack. A similar joint role as substrate binder and Lewis-acic catalyst has also been suggested for the zinc ion, which can activate alkaline phosphatase in place of magnesium.

Carboxypeptidase is yet another zinc-requiring enzyme. It consists of a single polypeptide chain of no less than 307 amino-acid residues, yet it requires a single zinc ion for activity, which surely illustrates the biological significance of this element. The zinc ion appears to play no structura role, since it can easily be removed, leaving an apo-enzyme which has identical optical activity and sedimentation pattern to the native molecul The zinc ion can be replaced by a number of dipositive, first-row transition-metal ions, but only four of these, Mn^{2+}, Fe^{2+}, Co^{2+}, and Ni^{2+}, give active complexes; the others, Cu^{2+}, together with Cd^{2+} and Hg^{2+}, produce only inactive species, though they bind to the same site. This lends support to the view that in this enzyme the Zn^{2+} acts as a Lewis acid (Fig.16), since the active ions are themselves all quite 'hard' with high Lewis acidity, whereas the remainder are distinctly 'soft' with only weak Lewis acidity.

Measurement of the stability constants of these seven metal–enzyme complexes suggested that the metal-binding site probably consisted of nitrogen and sulphur donors which seemed consistent with the known chemistry of zinc, but as if to illustrate the fallibility of this method, an X-ray crystallographic analysis has now been completed which clearly shows that the ligands are two histidine and one glutamate residues. This clearly shows the danger and limitations of deducing results from model complexes with regular coordination geometries, and applying them to enzymic sites which generally have a far from simple stereochemistry.

Carbonic anhydrase (carbonate hydro-lyase) is an extremely important zinc-potentiated enzyme found in considerable concentration in erythrocytes. It is responsible for catalysing the equilibrium:

$$CO_2 + H_2O \rightleftarrows HCO_3^- + H^+,$$

Carboxypeptidase

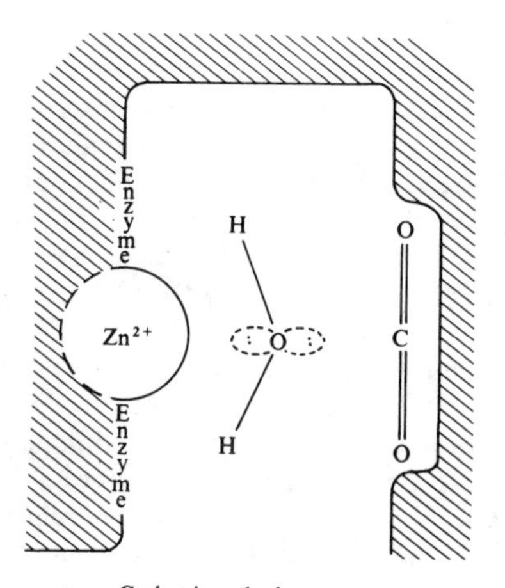

Carbonic anhydrase

FIG.16. A suggested role for zinc in carboxypeptidase and in carbonic anhydrase. In carboxypeptidase the binding of the substrate to the zinc ion results in a polarization which facilitates hydrolysis, whereas in carbonic anhydrase binding of water (a substrate) simply holds it in the correct stereochemical position.

which in turn provides much of the buffering capacity by which the pH of mammalian blood is controlled. Metal ions other than Zn^{2+} are able to bind to the apo-enzyme, but only Co^{2+} restores any activity. From magnetic susceptibility and spectrophotometric studies on the Co^{2+}–enzyme complex it has been concluded that the binding site is tetrahedral. Moreover, the visible spectrum of the complex remains unaltered even when carbon dioxide was bound, indicating that this substrate is not bound directly to the metal. Fig.16 shows a possible role for zinc in the enzyme. It is important to realize that this suggested role differs from that suggested for metal ions in almost every other enzymic reaction. Normally the zinc coordinates with the oxygen atom of a phosphate or carboxylate group, functioning by withdrawing electrons, thus encouraging nucleophilic attacks on adjacent bonds. In this case the role is different to that of zinc in all other enzymes, since it appears to act not as a Lewis acid but by ensuring the presence of a water molecule in the correct stereochemical arrangement for nucleophilic attack on the carbon dioxide.

Cadmium and mercury toxicity

In vivid contrast to zinc, neither mercury nor cadmium have any know biological role, indeed, they are both highly toxic. The toxicity varies con siderably with species and molluscs seem particularly resistant to cadmiur poisoning. Surprisingly, even though it does not appear to have a functior role, they accumulate the element, and samples taken from below the outfall of a zinc smelter on the Severn estuary have been shown to contai up to 1000 p.p.m. of cadmium in dried tissue.

Cadmium poisoning is fortunately quite rare, though isolated cases do occur from such unlikely sources as the contamination of food cooked in earthenware vessels containing yellow cadmium sulphide as a pigment in the glaze. Recently cadmium has been detected in mammalian kidney as a ternary Cd–Zn–protein complex containing up to 3 per cent cadmium. This protein, metallothionen, has a molecular weight of 10 000 and an unusually high proportion of cysteine residues, about one-third, which make it ideally suited for binding the Group IIB metals. The function of the complex is still unknown, but it is quite possible that it forms part of a detoxification pathway. Such a mechanism may be necessary, particularly for the male, since even small doses of cadmium damage the seminiferous tubules and kill the spermatozoa. A number of other specific effects of sub-lethal cadmium toxicity are known and in particular it has been shown to interfere with iron metabolism, producing symptoms similar to iron deficiency.

Poisoning by mercury has long been known. The phrase 'as mad as a hatter' refers to the fate of hatters who used mercury(II) nitrate to treat furs. Mercury in most forms is easily absorbed, but only relatively slowly eliminated, and therefore the effects of exposure are cumulative. Mercury compounds are usually considered to be more toxic than cadmium ones, though comparative figures for lethal dosage are rare. However, it is certain that mercury compounds are potentially more hazardous, principally because of their volatility. Elemental mercury is of course a liquid and moreover has a high vapour pressure compared with other metals (about 0·001 mm at room temperature), which means that the mercury concentration in the atmosphere can rise to 2 p.p.m., which, though seemingly quite low, is over 200 times the recommended maximum permitted concentration. Even more hazardous are volatile organometallic compounds, such as dimethyl mercury, with boiling points which are much lower than that of the metal itself.

Poisoning by mercury is more common than cadmium poisoning mainly because of the greater use made of mercury compounds. The use of mercurial ointments once widely recommended for treatment of syphilis has been discontinued since the 'cure' would seem to have been as deadly as the disease, but alkyl mercury(II) halides (R–Hg–X) continue to find use as powerful diuretics, albeit under the most strictly controlled conditions. They are given in the form of intra-muscular injection since they are too toxic to be given intravenously and are poorly absorbed orally. The most widespread use of mercurials is as fungicides and bactericides though this is now being actively discouraged. Nevertheless the usage of compounds such as phenyl mercury(II) acetate in this role persists and contamination by such materials, together with industrial wastes, form the major source of mercury pollution.

Mercury residues are found in almost all living organisms but there is now an alarming body of data showing that the mercury is gradually concentrated in organisms which are high in their food chains. Prior to this it was widely believed that the end-product of the breakdown of all mercurials would be an insoluble immobile salt. Clearly, the organomercurials are more persistent than was thought hitherto, and there is also convincing evidence that inorganic mercury may be converted to the much more mobile, and hence more dangerous, dimethyl mercury by some strains of anaerobic bacteria found in eutrophic waters. The toxic action of both mercury and cadmium has been widely studied and has been related to their proclivity for binding thiol groups. This can clearly account for their interference with the reactions of the Krebs cycle, where coenzyme A, containing a functional thiol group, is of great importance. Not only do these metals bind the thiol form of the coenzyme, thus rendering it useless, but they have also been shown to catalyse the hydrolysis of thiol-esters, thus breaking down reactive intermediates. The driving force for this reaction comes once again from the enormous affinity of these metals for sulphur ligands. The mechanism of toxicity seems quite general, and the diuretics mentioned previously are also believed to work by forming mercury derivatives of important biological thiols.

5. The metals of group IIIB and beyond

The chemistry of the later main-group metals

THE chemical property common to all the metals discussed so far, and of course the basis of the definition of a metal given in the first chapter, has been their propensity to lose all their valence electrons to form a cation. However, the removal of each successive electron grows progressively more difficult so that the total energy required to abstract several electrons increases very rapidly indeed. (Compare the first ionization energy with the sum of the first three ionization energies and so on for any of the elements of Group III and beyond.)† Consequently, it is apparent that the formation of a cation in the group valency or oxidation state will eventually become energetically prohibitive, and with this in mind the explanation of two related trends becomes evident.‡

First, it is noticeable that on moving across the periodic table from Group IIB to Group VIB, the lighter elements of each group, which have the largest ionization energies, are increasingly likely to form their compounds either by covalent bonding or ultimately even by directly accepting electrons to become anions. As a corollary therefore, cation formation, and hence metallic classification, is increasingly limited to the heavier members of each group. Thus in Group IIIB only the lightest member, boron, is a non-metal, and even boron is metalloid, but by the time Group VIB is reached, only the heaviest member, polonium, has any significant metallic properties.

Secondly, even for the heavier members of each group, whose ionization energies are somewhat smaller than those of the lighter congeners, abstraction of all the valence electrons is still energetically expensive, and as a result it is not uncommon for them to form cations of lower charge, for example, Tl^{+}, Pb^{2+}, or Bi^{3+}, by losing only part of their complement of valence electrons. In each case this leaves a cation with an outer electron configuration ns^2 and these two s-electrons which may then be removed only with difficulty are commonly described as the 'inert pair'.

In summary then it can be appreciated that the later main-group elements are unlike the alkali metals or alkaline earths in that they may form cations of more than one valency state. The lowest oxidation state represents removal of the p-electrons of the valence set, the higher represents removal of all

† See Pass: *Ions in solution (3): inorganic properties* (OCS 7).

‡ The group oxidation state is equal to the sum of all the valency electrons.

valence electrons, thus the two states always differ by two, for example, Tl^{III}/Tl^{I}, Pb^{IV}/Pb^{II}.

Taking the elements of Group IIIB as an example, then, on the basis of previous argument it would be reasonable to predict that they should all form compounds in the group oxidation state of +3, and indeed this is largely the case. However, the 'inert pair' effect could be important since the first ionization energy is very much less than the second and subsequent steps and this suggests that univalency might be feasible. In the case of boron this is not in fact so, as the energy gained by creating three covalent bonds outweighs that to be gained from a single ionic one; but on moving further down the group the balance of energies changes and the univalent state becomes increasingly important. Both Al^+ and Ga^+ are known, but only in melts, whilst In^+ and Tl^+ are both known in aqueous solution. Indeed, thallium(I) is usually more stable than thallium(III), and the redox properties of this couple dominate the aqueous chemistry of this element, though the value of the redox potential varies considerably both with the pH and the nature of the anions present in the acid solution.

$$Tl^{(III)}{}_{(aq)} + 2e^- \longrightarrow Tl^{(I)}{}_{(aq)}, \quad E^{\ominus} = +1{\cdot}26 \text{ V}.$$

Chemically Tl^+ has many similarities to both potassium and silver ions, not least because of the similarity in their ionic radii– ($K^+ = 0{\cdot}133$ nm, $Ag^+ = 0{\cdot}113$ nm, $Tl^+ = 0{\cdot}149$ nm). Like silver (I), thallium(I) is a relatively 'soft' cation and so differs from potassium by having a moderate affinity for sulphur ligands.

Whilst boron does not form a discrete B^+ cation, neither does it form any compounds in which the trivalent B^{3+} cation may be recognized, and examining the chemistry of this group in more detail, it is easy to see why this should be. For boron, the total of the first three ionization energies far outweighs any possible compensating hydration or lattice energies, so that no discrete B^{3+} ion is ever formed. Echoing the arguments used to rationalize the chemistry of beryllium, it is also possible to take the view that the hypothetical B^{3+} ion, with an estimated radius of 0·02 nm and hence a massive charge/radius ratio (estimated at 15). would be so powerfully polarizing as to make covalency inevitable. Indeed, no simple aquo-ion $[B^{III}(OH_2)n]^{3+}$ has ever been detected, but boric acid, H_3BO_3 (which is more usefully written as $[B(OH)_3]$), is well known.

A count of the valence electrons associated with boron compounds of this type of compound reveals that they are formally 'electron-deficient'. Not surprisingly then, they readily act as electron acceptors or Lewis acids. Boric acid, for example, is not a simple protonic acid but instead reacts by abstracting hydroxide ion to form the borate anion:

$$H^+ + [B(OH)_2O]^- \not\leftarrow [B(OH)_3] \xrightarrow[pK_a\ 9{\cdot}2]{OH^-} [B(OH)_4]^-.$$

A count of valence electrons in this ion shows that boron now satisfies the eight-electron rule. Such behaviour is not limited to boron, since, if they are considered to be covalent, certain aluminium compounds such as the chloride, $AlCl_3$, are also electron deficient. Such compounds are Lewis acids, and aluminium chloride in particular is used in this role as a catalyst in organic syntheses, notably in the Friedel–Crafts reaction. However, as such compounds are unstable in water and therefore not likely to be directly important biochemically, they will not be discussed further here.

Whilst in a non-aqueous environment aluminium resembles boron quite closely, the properties of the compounds of the two elements differ quite strongly when water is present. Thus although aluminium chloride, $AlCl_3$, is a low melting-point solid (m.p. = 193 °C), and this can be regarded as evidence favouring covalency, in aqueous solution it rapidly dissociates to produce the aquated aluminium ion. The resulting solution is acidic since the polarizing power of the small highly-charged Al^{3+} ion is sufficient to induce hydrolysis, and one result of this is that aluminium salts of weak acids such as the sulphide or carbonate cannot exist in contact with water.

$$[Al^{(III)}(OH_2)_6]^{3+} \xrightarrow[pK_a\ 4{\cdot}98]{} [Al^{(III)}(OH_2)_5(OH)]^{2+} + H^+.$$

Given the high charge on the aluminium ion, it is hardly surprising that it forms so many insoluble salts. Equally, a range of stable aluminium complexes are known, particularly with anionic oxygen donor ligands. For example, the reaction of excess oxalic acid with aluminium salts produces not the simple oxalate salt but instead the complex trisoxalatoaluminate(III)anion $[Al^{(III)}(C_2O_4)_3]^{3-}$. The three remaining members of this group – gallium, indium, and thallium – also form well-defined series of trivalent salts which, like their aluminium analogues, dissolve in water to give the aquated trivalent metal ion. These ions are also subject to extensive hydrolysis which, contrary to expectations, *increases* down the group, so that $[Tl^{III}(OH)_6]^{3+}$ is stable only below pH 1·5, and at higher pHs it readily hydrolyses to form hydroxides or hydrated oxides. This and other irregularities in the group-trends arise because the ions Ga^{3+}, In^{3+}, and Tl^{3+} do not have the 'noble-gas' configuration but instead are like the metals of Group IIB and have an outer set of filled d-orbitals. As these particular electrons have only a poor capacity to shield the nuclear charge, the result is a higher *effective* ionic charge, which offsets the increase in atomic radius.

The chemistry of the remaining main-group elements can be summarized in an extension of the pattern shown by the Group IIIB elements. Moving to Group IVB, it is now the first two elements, carbon and silicon, which are non-metals; the third, germanium, is metalloid and only the heaviest two, tin and lead, are strictly metallic. In each case they form compounds in two oxidation states, much as expected, and the stability of the highest (+4) oxidation level decreases down the group. Thus tin(II) acts as a reducing agent, whilst lead (IV) compounds are often oxidizing. Each metal forms typically ionic compounds when in the lower oxidation state, but compounds in which they adopt the higher oxidation state not unnaturally have considerable covalent character. This is reflected by the range of organometallic compounds (compounds with direct metal-carbon bonds) formed by germanium, tin, and lead. Such compounds (which incidentally form an interesting exception to the original definition of a 'metal'), are relatively stable even in aqueous solution, though the exact properties depend on the nature of the ligands.

The elements of the remaining groups show an increase in non-metallic character. In Group VB, antimony and bismuth are usually considered to be metals, though in many of their properties they have considerable similarity to the earlier non-metallic members of the group. In Group VIB only polonium is at all metallic, and since it is both rare and unstable in all its isotopic forms it need not be discussed further.

Methods for studying the later main-group metals *in vivo*

The concentration of aluminium in biological samples is readily determined by a variety of classical analyses, but the detailed study of aluminium biochemistry has been hindered by the absence of any isotope with half-life greater than one second. Reported studies on gallium and indium are few, but thallium has been studied more intensively in its guise of a probe for potassium. The two isotopes, ^{203}Tl and ^{205}Tl, that make up more than seven-tenths of natural thallium both have non-zero nuclear spins and may be studied by n.m.r. spectroscopy. In addition ^{202}Tl (electron capture; half-life 12 d) and ^{204}Tl (a β^- emitter; half-life 3·8 years) make useful tracers.

The metallic elements of Group IVB are relatively amenable to study, with a number of isotopes suitable for tracer work including ^{113}Sn (electron capture; half life 120 d) or, particularly important, the much-employed ^{210}Pb (a β^- emitter; half-life 20·4 years). In addition tin compounds may be examined by the use of Mössbauer spectroscopy using ^{119}Sn, though their biochemical insignificance has limited the application of this method. Similarly the remaining main-group metals are amenable to study by a variety of means, but reports are few on account of their limited biological role.

Terrestrial distribution

The geochemistry of the Group IIIB elements is relatively more complicated than that of other groups met so far, reflecting the larger range of properties exhibited by this group. The non-metal boron is rare but is found locally in high concentrations in 'borate beds', the evaporates of waters which have percolated through boron-containing rocks and soil. Aluminium, on the other hand, behaves as a typical metal. It is the most abundant metallic element in the earth's crust, with an average concentration of over 7 per cent. As would be predicted from a brief review of its chemistry it always occurs as the Al^{3+} ion, chiefly in the alumino-silicate minerals mentioned previously in Chapter 2.

Less frequently, aluminium is to be found in simple oxides or hydrated oxides, particularly bauxite, reflecting its affinity for oxygen. The actual form of these deposits depends on the geological influences which have acted on them. Gallium and indium are usually found as minor components of aluminium ores, but even the richest of these contain less than 1 per cent of these metals. By way of contrast, thallium is found not as Tl^{3+} but as Tl^{+}, and not associated with oxide ores but with zinc and iron sulphides, and so is classified as chalocphile rather than lithophile, distinguishing it from the remainder of the group. High concentrations of the heaviest members of this group are uncommon and biologically insignificant.

In Group IVB, germanium is extremely widespread, though only of moderate abundance, typically 1 - 5 p.p.m. The intermediate value of its charge/radius ratio means that it can be included in most other minerals without disrupting their structure and so there are few distinctive germanium-containing ores. In complete contrast, tin is found almost exclusively as the complex tin (IV) oxide, cassiterite; and lead, by way of further distinction, is strongly chalcophile, being found chiefly in galena (PbS). In this form it occurs in high local concentrations which can be important biologically. Antimony and bismuth are also found as sulphides but apart from limited mineralization they are both rare, with typical levels for both being less than 1 p.p.m. Finally, polonium, all of whose isotopes are radioactive, occurs at such vanishingly small levels (2×10^{-10} p.p.m.) that it may be ignored.

The later main-group metals *in vivo*

Despite being the most abundant of all the metals in the earth's crust, aluminium has no obvious biological role. This statement should perhaps serve as a timely reminder of the many gaps between observation and explanation in the biochemistry of the metals. Though convincing reasons for this surprising lack of functional role are still missing, two possible contributory factors should be considered. First is the low availability of aluminium in the alumino-silicate minerals, though despite this, the aluminium level in fresh-water (0·24 p.p.m.) is still higher than that of many other 'functional' metals, so that this may be only a part of the whole explanation. The second factor is the insolubility of many aluminium salts, particularly the phosphates. Given the importance of polyphosphates in bio-energetics it seems likely that aluminium could have been excluded because of its interference with their metabolism. This argument by itself, however, is less than convincing since calcium phosphates are equally insoluble, though in contradiction this is the very property which recommends their employment. Perhaps then it has been the evolutionary selection of calcium that has precluded any similar use of aluminium.

The published data for the aluminium content of both plant and animal tissues show that levels are more variable than consistent. In animals, levels of 1 p.p.m. are rarely exceeded except in the lungs, where the aluminium content may reach 80 p.p.m., presumably because of the high aluminium content of airborne dust particles.

The effect of aluminium on plants has been widely studied, with some confusing results. Most plants seem to contain aluminium, but excess is toxic. Experiments on excised roots have shown that Al^{3+} ions can both increase and decrease the uptake of other metal ions, depending on the external experimental conditions. The uptake of aluminium itself is dependent mainly on the pH of the soil, being best in slightly acid soils but less under alkaline conditions, owing to the formation of insoluble 'aluminium hydroxide' polymers. It is probable that the antagonism between calcium and aluminium uptake is a result of increased soil pH caused by adding lime, rather than direct competition between the two ions. An interesting consequence of this is the colour of hydrangeas. In acid soil hydrangeas may be made to turn blue by feeding aluminium salts, but in calcareous soils they remain pink.

Aluminium salts find some uses medicinally. Their acid nature makes them suitable for use in astringent lotions and mild antibacterial washes and, because of the low toxicity of aluminium, aluminium hydroxide is much used as an antacid in proprietary preparations. Extended doses can be harmful though, because precipitates of aluminium phosphate, which are formed in the alkaline conditions of the small intestine, can lead to phosphorus deficiency. It has been used for just this purpose to treat patients with insoluble phosphate stones.

Aluminium hydroxide is considered an ideal antacid since it exhibits a buffering effect and establishes a pH of about 4. Neutralization proceeds in a series of steps. In strong acid solution the $[Al^{(III)}(OH_2)_6]^{3+}$ ion will be formed but, as described earlier in this chapter, this cation is a weak acid, pK_a 4·9, so, as the pH of the solution rises it undergoes self-ionization, which increases the pH. The balance of these two effects gives an equilibrium about pH 4·0.

Not only do plants normally contain a higher level of aluminium than animals but the range within the plants themselves is also quite large. A few species are aluminium accumulators and these have typically yellowish leaves and blue fruits. The coloration is caused by the formation of a coloured 'lake' and a similar phenomenon is responsible for the blue colour in hydrangeas and certain other flowers. Amongst the seed plants, *symplocos* contains up to 3 per cent aluminium and the club-mosses form a group of aluminium-accumulators. Well known in this respect is *Lycopodium tristachya*, which contains ~ 1 per cent aluminium.

None of the three remaining elements has any known biological role, and all are toxic. Gallium and indium are only moderately so, but thallium is extremely toxic, and in at least one case has been used to commit murder. Information on the mechanism of thallium toxicity is scarce, but it is most likely due to some interference with sodium-potassium metabolism.

In Group IVB: germanium has no known biological role; the status of tin is still uncertain (though there are preliminary indications that it may be essential); and lead is distinctly toxic. The levels of tin in most soft tissues range from about 0·1 p.p.m., rising to 1 p.p.m. in fresh bone, but there are interesting geographic variations and the mean levels in kidney and liver are lower for peoples in underdeveloped areas of the world than for similar sample groups in the urban Western countries. This no doubt reflects the greater usage of convenience-foods in the Western world since large amounts of tin can accumulate in canned foods, particularly if the tin plate is not lacquered. However, toxicity is unlikely and mice given 5 p.p.m. of tin in the drinking water over their whole lifespan exhibit no ill-effects.

Lead is widespread in the lithosphere, with part at least of the lead in the soil being easily extractable. Not surprisingly, therefore, lead can be detected in all plant materials, with levels found to be roughly proportional to the amount of extractable lead in the soil. In regions where there are high levels of lead in the soil, for example, around lead mineralizations or some industrial sites, plant levels may be 10 - 50 times those found elsewhere, though the plants are not true accumulator species. However, in such areas the vegetation does show selection for specific tolerance to high levels of available lead, and indicator plants such as *Amorpha canescens* are not uncommon. The levels of lead in urban vegetation are similarly variable, though mean levels are in general higher than those of rural samples. However, much of the increase can be traced to airborne pollutants, particularly particulate matter from car exhausts, deposited on the exterior of the plant. In a typical experiment, vegetation at the side of a busy road was found to contain 250 p.p.m. of lead, whilst vegetation collected 50 m from the road contained less than 50 p.p.m. lead; but even higher levels of contamination (up to a hundred-fold increase) have been exceptionally reported.

The lead burden of grazing animals is not unnaturally related to the lead levels of the pasture land, but a number of other factors combine to alter and obscure this simple relationship. It is an interesting but unexplained observation that, with the exception of a few root-crops (notably carrots), absorbed lead from the soil is confined mainly to the plant roots. Little is translocated to the shoots and leaves, so that only a small fraction of the soil lead actually reaches the animal via the fodder. This is modified by two further factors: first, airborne pollutants externally deposited on the plants can increase the effective lead available, and secondly the level of internal plant lead shows strong seasonal variations, being least during periods of active growth and greatest just after inactivity. Finally, the ingested lead is very poorly absorbed by ruminants (and by non-ruminants and typical experiments suggest that only about 1 per cent of the dietary intake is retained.

For man, this picture is even more complex. Despite popular belief, lead from exhaust fumes does not always (nor even often) constitute the prime source of lead for urban man. Instead, food and water provide the majority of the daily intake. Estimates vary of course, but it has been suggested that on average an adult will ingest some 300 μg of lead daily and inhale about 15 μg in the same period. Nevertheless, lead pollution from motor-vehicle exhausts is not negligible, since a certain amount of the lead in food must originate from exhaust emission; but to demonstrate the difficulty of correlating blood-levels of lead with the degree of urbanization of the environment, it has been shown that natives in New Guinea have levels similar to those found in natives of Los Angeles. However, specific exceptions are important, and the alarmingly high levels found in Tokyo's traffic police do point to the inherent dangers.

In general, however, the levels reported today do not differ greatly from those reported over 30 years ago, and it has been suggested that lead from traffic fumes has replaced lead from other sources, particularly that from drinking-water carried in lead pipes, as the principal but unwanted additional source of lead. Indeed the increase in lead in urban vegetation over the last 50 years can be traced quite accurately by measuring the lead content of each annual ring of urban and rural trees. The urban results show a progressive increase in the lead content of the younger rings that is not matched by that in rural trees. Similar results, though with very much smaller lead levels, have been reported by analysing the permanent arctic snows. Here, the deeper, older levels show significantly lower levels of lead (20 p.p.b.) than upper, more recent levels (200 p.p.b.).

The toxicology of lead is well documented and, because of the pollution problem, particular attention has been paid to the cumulative effect of relatively low doses. The major proportion of ingested lead is rapidly excreted, but of course as the dose increases relatively more is absorbed. Much of this is immobilized and rendered less harmful by incorporation in bone and hair, but some is concentrated in the liver, with deleterious effect. Apart from a rise in levels of lead in the blood above the 'normal' range (typically 30 μg per 100 cm^3 blood) clinically recognizable lead poisoning is evidenced by an increase in the excretion of δ-aminolaevulinic acid, a porphyrin precursor; which no doubt accounts for the lead-induced anaemia. This excretion has been traced back to the inhibition of the synthesis in erythrocytes of a specific enzyme, aminolaevulinic acid dehydrogenase; the quantity of enzyme being inversely proportional to the blood-lead level.

Cases of chronic lead poisoning are also characterized by neurological symptoms, and on the basis of this and high blood-lead levels it has been suggested that there may be a link between lead pollution and personality disorders, particularly in juveniles – though the case is not yet proved. Treatment of lead poisoning is fairly simple and relies on the chemical affinity of lead for thiol ligands such as British anti-Lewisite (B.A.L.; $HSCH_2 \cdot CH(SH) \cdot CH_2OH$).

6. The transition metals

The chemistry of the transition metals

THE transition metals can be loosly defined as those elements whose ions have a partially filled set of d-orbitals.† Considering, then, that this definition encompasses over 20 elements, each with a rich and diverse chemistry it is hardly surprising that useful generalizations about their behaviour are few, and it must be conceded that deductions based solely on electronic configurations and simple parameters such as ionization energies and hydr tion energies give only the poorest guide to the aqueous chemistry of mos transition metals. Nor is it possible to discern many significant group-trends, since the smaller members of the first row differ markedly from their heavier congeners in the second and third rows, though it is noticeab that for each triad the two larger elements, which happen to be similar in size, also have similar chemistries. It is fortunate, therefore, that with the exception of molybdenum, the biologically significant transition metals all seem to be members of the first row, so that it is possible to restrict much of the subsequent discussion to just these elements.

Unlike the s- and p-block metals, which form ions in only one, or at most two, oxidation states, the d-block transition metals are able to exist in a variety of oxidation states, both positive and negative.‡ The chemistr of the cations is dominated by their ability to act as Lewis acids, forming stable complexes with a wide variety of Lewis bases.§ In many cases the interaction between the ligand(s) and the metal favours one particular oxidation state of the metal at the expense of the others.

With ligands of biological significance, which tend to have nitrogen-, oxygen-, or sulphur-containing donor groups, it is generally true that, with the exception of titanium, either the +2 or +3 oxidation states are the mo stable for all the first-row elements. Of course it is perfectly possible to obtain most of these elements in higher oxidation states, particularly with oxide or fluoride ligands, but such species as $[Cr_2^{VI}O_7]^{2-}$ or $[Mn^{VII}O_4]^-$ tend to be strong oxidizing agents and, under biological conditions – that is to say, in an aqueous environment – they usually react fairly rapidly

† Strictly speaking, the lanthanides and actinides with partially filled f-shells are also described as transition elements, but since they do not seem to have any significant biological role they will not be discussed further here.

‡ With so-called π-acceptor or π-acid ligands many transition metals are able to for complexes in which the metal is formally of zero oxidation state or lower, and th theoretical interpretation of this is still a matter for discussion.

§ The term 'stable' is used here, not in the strict thermodynamic sense, but merely to imply that the complexes may be isolated without undue difficulty.

to give complexes which contain the metal ion in a lower oxidation state.†

In some cases the balance of energies is so fine that it is possible, under very similar chemical conditions, to obtain complexes which differ only in the oxidation state of the metal. For example, both $[Co^{II}(NH_3)_6]^{2+}$ and $[Co^{III}(NH_3)_6]^{3+}$ may readily be prepared in aqueous solution at room temperature. The cobalt(II) complex is readily oxidized to cobalt(III) by oxygen, and in this case it is possible to reverse the process by use of a suitable reducing agent. It is noticeable that the transition elements that are particularly involved in those biochemical processes such as electron transport or catalytic (enzymic) oxidations have this ability to move reversibly between different oxidation states.

A number of theories have been developed to explain the properties of transition-metal ions and their complexes and the simplest, which is quite successful, is known as the crystal field theory.‡ This makes the assumption that all ligands, whether anionic, e.g. the cyanide ion, CN^-, or merely dipolar, e.g. H_2O, can be treated simply as point negative charges and hence that bonding between the metal ion and the ligand is purely electrostatic. If the geometry of the complex is known, it then becomes a relatively simple matter to deduce the effect of the array of charges on the d-orbitals themselves.

In Fig.17 it can be seen that an octahedrally or a tetrahedrally disposed set of ligands splits the d-orbitals into two sets, differing by energy values defined as Δ_{oct} and Δ_{tet} respectively. Related splitting patterns can be derived almost as easily for other common coordination geometries. The splitting of the d-orbitals into non-degenerate sets in this fashion is a fundamental result of the crystal field theory and can be used to explain many of the commonly observed properties of transition-metal complexes.

Fig.17 further shows how, for an octahedral complex, it is possible to arrange six d-electrons in two extreme configurations, either $t_{2g}^4 e_g^2$ (high-spin, or spin-free) or $t_{2g}^6 e_g^0$ (low-spin, or spin-paired). Clearly, the latter arrangement is more stable since all the electrons are in lower-energy orbitals and the energy difference between the two configurations, the crystal field stabilization energy (CFSE), amounts to $-2\Delta_{oct}$ in this particular example. However, the CFSE is only one of the factors which determines the choice of spin state and it must be remembered that it requires the expenditure of energy to bring two electrons to pair up their spins. Therefore, only if the CFSE is more than the total spin-pairing energy will the ion adopt a lower-spin configuration, and if the CFSE is less than the pairing energy a high-spin configuration will result.

The actual magnitude of Δ, and therefore of the CFSE, depends ultimately on the nature of the ligands surrounding the metal but, surprisingly,

† See also Bell: *Principles and applications of metal chelation* (OCS 25)

‡ See also Earnshaw and Harrington: *The chemistry of the transition elements (OCS 13)*

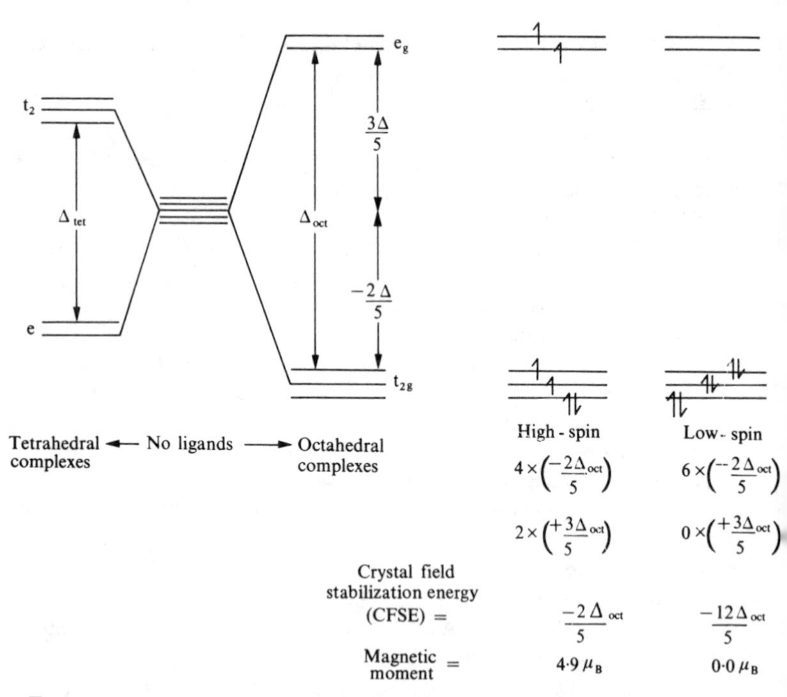

FIG.17. In the absence of ligands the five d-orbitals are degenerate, but on the formation of a complex this degeneracy is removed and the orbitals are split into two or more sets. Six d-electrons can be arranged in two extreme ways in an octahedral complex. The magnetic moment can be predicted using the 'spin-only' formula $\mu = \sqrt{N(N+2)}$, where N is the number of unpaired electrons.

the *relative* splittings produced by a whole series of ligands remain consta whatever the nature of the metal ion, so that most common ligands can b arranged in the 'spectrochemical series' which, in order of increasing ligand-field strength (Δ) is: $I^- < Br^- < Cl^- < F^- < OH^- < H_2O < NCS^- <$ py $< NH_3 <$ en $<$ 2,2-bipy $<$ *o*-phenanthroline $< CN^-$.

With the aid of these concepts it is now possible to interpret the magnetic properties of many transition-metal complexes (see Fig.17). Taking as an example the octahedral complexes of iron(II) (which has 6 d-electro it can be seen that the iron can have a maximum of 4 unpaired electrons, none. In the former case it will be paramagnetic, whereas in the latter it will be diamagnetic. The physical reality of this prediction is apparent by measuring the magnetic moment of two complexes: $[Fe^{II}(OH_2)_6]^{2+}$ whic has weak-field ligands, the water molecules; and $[Fe^{II}(CN)_6]^{4-}$, which ha strong-field cyanide-ion ligands. In the former the CFSE is less than the spin-pairing energy so that the complex is high-spin and paramagnetic wit a magnetic moment approaching 4·9 Bohr magnetons (μ_B), whilst the cyanide ligands produce a much greater CFSE so the latter complex is low-spin with no unpaired electrons, and is diamagnetic.

The CFSE is also involved when explaining why complexes of some ions, notably Cr^{3+}, Fe^{2+} (low-spin), and Co^{3+} (low-spin), undergo substitution reactions only slowly. It is noticeable that these kinetically inert ions are all subject to considerable stabilization in the octahedral ligand field. Therefore, reactions of their complexes – either by a dissociative (S_N1) mechanism, through a penta-coordinate intermediate, or by an associative (S_N2) mechanism, with a hepta-coordinate intermediate – result in the temporary loss of the CFSE in the intermediate, or transition state, so that the reaction is unfavourable.

Finally, the crystal field theory makes possible the understanding of the electronic absorption spectra of metal complexes. It is commonplace observation that most transition-metal complexes are coloured and experiment shows that the observed colour depends largely on the crystal field splitting of the d-orbitals, since the absorption of light provides the energy necessary for the promotion of a d-electron for one sub-set of d-orbitals to the other.

So far only the successes of the crystal field approach have been stressed, but unfortunately it suffers from the serious drawback that, whilst many of its predictions are qualitatively correct, quantitative calculations, with a few fortunate exceptions, are normally in error by several orders of magnitude. This is hardly surprising when it is recalled how simple are the assumptions on which the theory is based. For many purposes it has now been replaced by the use of molecular orbital theory, which takes into account the contribution of covalent bonding, and particularly π-bonding, to metal-ligand binding. Nevertheless, crystal field theory still retains much of its usefulness, at least in descriptive work, since the more rigorous molecular orbital treatment leads to exactly the same pattern of splitting and orbital occupancy as the simpler theory.

Little has been said so far about the shapes of transition-metal complexes except to assume an octahedral geometry for the purposes of the crystal field approximations. Other geometries are also common, with the regular tetrahedral and square-planar configurations being the most important, but many complexes have distorted or irregular geometries based on one of the three main types. Such distortions become increasingly common when polydentate ligands with their stricter steric requirements are used; they are important since they may result in considerable modification of the stability of any particular oxidation state.

In addition to the deviations from a regular geometry caused by steric effects from the ligand, the complexes may also become distorted by another effect, arising from the metal ion itself: the Jahn-Teller effect. Such a distortion will occur if the electronic configuration of the metal ion is such as to give rise to a degenerate state; i.e. one with more than one equivalent energy level. Although this sounds complicated, it is easy to recognize the operation of the Jahn-Teller effect in complexes of ions such as Cu^{2+}. Instead of six equivalent bonds, the so-called octahedral complexes of the Cu^{2+} ion generally have four short and two long bonds. Similar effects can be noted in complexes of some other ions and the more important cases will be mentioned. These ideas, along with others, will be developed in the subsequent discussion of the biochemistry of the first-row elements, but in view of their manifold differences it is more profitable to consider each element in turn rather than to trace the often tenuous link between them.

Methods for studying the transition metals *in vivo*

If the simplicity of the alkali-metal chemistry sets problems as to how their biochemical role can be studied, the very opposite is true of the transition metals, and there seems to be an almost embarrassing abundance of techniques for investigating their metabolic functions. This is certainly so for conventional 'wet' chemical analysis, where the distinctive redox properties of the transition metal ions and their disposition to complex-formation has encouraged the development of a wide variety of individual methods for the determination of the trace amount of these metals in biological samples. However, whilst analysis is essential, it is unfortunately true that chemical methods are mostly destructive and since the chemistry and hence the biochemistry of the transition metals depend intimately on the nature of the ligands to which they are bound, techniques that can directly yield information about the intact complexes will clearly be more valuable.

As with most other metals, radioactive tracer studies are particularly valuable in determining metabolic pathways and particular examples are quoted later in this chapter. The understanding of transition-metal biochemistry also owes much to a number of important structure determinations (particularly those of the iron-porphyrin complexes) by means of X-ray crystallography, but the bulk of the present information on the subject has come from two different but complementary methods – electronic absorption spectroscopy and electron spin resonance (e.s.r.) spectroscopy.

As has already been indicated, the characteristic colours of transition-metal complexes are caused by the absorption of light with the appropriate energy to induce d-d transitions between the sets of orbitals split by the crystal (or ligand) field. Thus the absorption is directly related to the nature of the ligands bound to the metal. Hence by an empirical comparison with complexes of known structure, assisted, if possible, by the application of the somewhat under-developed theory, it may be possible in favourable cases to deduce a good deal about both the nature of the donor atoms and also the stereochemistry of the complex. It is only fair to point out that there are sometimes quite important discrepancies between the structural information derived from spectroscopic data and that derived by more direct means, such as X-ray crystallography.

These discrepancies, and other problems which arise in the interpretation of spectroscopic data reflect the fact that many of the biologically important transitition-metal complexes (metallo-enzymes and the like) have 'irregular' stereochemistries. Expressed more realistically, it has become increasingly apparent that they have stereochemistries which do not conform to the simple octahedral, tetrahedral, or square-planar models upon which present interpretations are based, with the result that an understanding of the spectroscopic properties of species such as the copper-potentiated enzymes described later in the chapter has actually required a substantial advance on the simple theories described here.

Many of these comments are also applicable to studies on the magnetic properties of the biologically important transition-metal complexes. The determination of bulk magnetic susceptibility can give an indication both of oxidation state and also of electron distribution; but once again a multitude of effects associated with both the magnitude of the ligand field and the detailed stereochemistry of the complex conspire to make the simple spin-only interpretation suspect.

A more detailed picture, but one which requires increasingly sophisticated theoretical treatment, can be derived from e.s.r. spectroscopy. This involves the examination of the behaviour of unpaired electron(s) when irradiated by a microwave source in the presence of a strong external magnetic field. This method is particularly sensitive to changes in the nature of the ligands but despite many studies of simple 'regular' complexes, interpretation of data on metallo-enzymes is still difficult and, indeed, has once more provided a stimulus for new advances in theoretical treatment.

The terrestrial distribution of the d-block transition elements

Fig. 4 on p. 12 shows that the distribution of the first-row transition elements forms a peak on the major exponential trend, with iron having the highest abundance. Despite their substantial abundance these elements are relatively scarce in the earth's crust, the reason being that they are generally siderophiles or chalcophiles and are therefore concentrated in the core or mantle. Nevertheless, there remain traces of each of these elements in the rocks and soil of the earth's surface and these elements divide into two classes: those with average concentration of less than 100 p.p.m., that is to say scandium, vanadium, chromium, cobalt, nickel, and copper; and those with concentrations one or more order of magnitude greater, namely titanium, manganese, and iron. Whilst the availability of each of these elements is likely to be linked to their overall concentration, numerous other factors may modify this. In the case of scandium, for example, the principal chemical form is the Sc^{3+} ion which has geochemical characteristics similar to aluminium and the Al^{3+} ion; so that Sc^{3+} is largely bound up in the silicate rocks and is hence unavailable.

As the utilization of any one of these elements depends initially on its absorption by plants, which in turn usually implies transport out of the soil in aqueous solution, we might expect that concentration of each element in ground and natural waters could be used to give a measure of availability. One factor which has immediate influence is the soil pH, since under alkaline conditions all the first-row metals form insoluble hydroxides, which reduce their availability. However, it seems that chemical characteristics are just as important as availability; molybdenum, the only heavier transition metal of biological importance (required by many bacteria), has only a very low concentration in rocks and soil, (average less than 2 p.p.m.) and in fresh-water it has a lower concentration than any of the first-row elements.

The transition elements *in vivo*

With the possible exception of iron, all the biologically important transition elements are quite precisely described by the older designation of 'trace elements' and all, including iron, are certainly micronutrients. Thus, taking man as an example, the average human being contains amounts ranging from a few milligrams (for chromium) to several grams (for iron), but in every case the internal levels are finely controlled on daily intakes of just a few milligrams, which in many cases is a substantial over-supply. In man the omnivorous diet usually ensures that deficiency diseases associated with lack of the transition elements are uncommon (one exception to this rule is the anaemia associated with iron deficiency), whilst toxicities are equally rare. Particularly interesting are the many interactions between the micronutrient elements, of which anaemias induced by cobalt or copper deficiency are perhaps the best-known cases. Examples of these interactions, together with further discussion of the roles of individual elements, are given in more detail later.

Unlike animals, plants, which extract their nutrients from a limited area, often suffer from marginal deficiencies and respond rapidly to supplementation; conversely many other examples of distinct toxicities due to high local concentration of a particular element have been documented, though, unlike deficiencies, these are difficult if not impossible to treat on a large scale.

The functions of the biochemically significant transition metals can be roughly divided into catalysis and transport. Catalysis in turn can be classified as Lewis acid or redox, and overlapping with this latter category is the function of electron transport. Another equally important transport function, particularly associated with the iron-porphyrins, is that of oxygen transport in mammals, which is described in more detail later.

Scandium ($3d^1$ $4s^2$)

Although its electronic configuration formally places scandium at the head of the first transition series, it provides the exception to most of the statements made previously, since its aqueous chemistry can be understood simply by considering the ionization and hydration energies. The balance between these two factors ensures that, apart from the metal itself, scandium exists only in the +3 oxidation state. As the Sc^{3+} ion has no d-electrons and a closed-shell configuration, it is both colourless and diamagnetic and hence it is more appropriately considered as a congener of aluminium than as a transition-metal ion. Despite the difference in ionic radii (Sc^{3+} : 0·068 nm; Al^{3+}: 0·045 nm) the aqueous chemistries of the two ions are generally rather similar.

Scandium is widely distributed in the geosphere, though only at low concentrations, and it has been found, at negligible levels, in animal tissue (10 p.p.b. in mammalian heart) and somewhat higher concentrations in plants. However, as plants mostly contain aluminium this is hardly surprising. No evidence of deficiency disorders nor symptoms of significant toxicity have been recorded and, up until the present, no functional role has been ascribed to this element.

Titanium ($3d^2$ $4s^2$)

Surprisingly, most compounds of titanium contain the metal formally in the +4 oxidation state, though an inspection of the ionization potentials indicates that the amount of energy required to remove all four electrons is prohibitively high. The explanation is simple; most compounds of tetravalent titanium are covalent in character ($TiCl_4$ is a colourless *liquid*) and bear a close resemblance to similar compounds of tin(IV). Titanium tetrachloride is readily hydrolysed and the resulting solution is strongly acid which can be accounted for by the ready deprotonation of water molecules bound to the metal owing to the high charge/radius ratio of the putative Ti^{4+} ion:

$$[Ti^{IV}(OH_2)_6]^{4+} \rightarrow [Ti^{IV}(OH_2)_4(OH)_2]^{2+} + 2H^+ .$$

The addition of further base rapidly produces a colloidal precipitate of the oxide rather than the formation of the expected hydroxide.

Titanium(IV) is readily reduced by zinc in dilute acid to give titanium(III), which forms the well-known purple aquo-ion $[Ti^{III}(OH_2)_6]^{3+}$. In this oxidation state the metal can be considered as a typical transition metal since the colour of its complexes arises from the promotion of the single d-electron. Titanium(III) is paramagnetic with a magnetic moment about $1.7\mu_B$, close to the spin-only value for a single unpaired electron, and its complexes are mildly reducing since the metal readily reverts to the +4 oxidation state. Compounds of titanium in other oxidation states, +2, 0, -1, are all known, but are unstable in the presence of water.

Titanium is quite abundant throughout the earth's crust, but no definite role has yet been ascribed to it. It may well have a functional significance – certain marine organisms, such as plankton, have an unexplained ability to accumulate this element. For example, the dinoflagellate *Gymnodinium brevis* contains up to 80 p.p.m. of titanium, which represents a concentration factor of at least 8×10^4, since the level of titanium in sea-water does not normally exceed 1×10^{-3} p.p.m., owing to the insolubility of the oxide. The chemical state of titanium in these organisms is unknown, as are its functions (if any), but judging from its chemistry it could be involved in some redox process.

Vanadium ($3d^3 4s^2$)

Only the most electronegative elements, such as fluorine and oxygen, form compounds with vanadium in which the metal adopts its highest (+5) oxidation state. Naturally, such compounds have considerable covalent character so that it is not possible to identify any discrete V^{5+} species. Vanadium pentafluoride, for example, is a colourless, diamagnetic liquid, boiling at 48°C. The oxide, V_2O_5, is also diamagnetic, although it is not colourless but bright orange. However, d-d transitions are certainly not involved.

By virtue of the *formal* d^0-configuration of the vanadium, V_2O_5 bears some resemblance to P_2O_5 (P_4O_{10}), though there are a considerable number of detailed differences in the chemistry of the two compounds. The similarity is most marked in the reaction of the oxides with base to give vanadate $(VO_4)^{3-}$ and phosphate $(PO_4)^{3-}$ salts respectively. However, the monomeric $(VO_4)^{3-}$ anion is stable only in quite strongly alkaline solution, pH $\geqslant$ 13 and below this it rapidly aggregates to form polymeric oxyanions reminiscent of, but stoichiometrically different from, the polyphosphates. In acid solution these polyvanadates decompose to give the monomeric VO_2^+ cation. This is a powerful oxidant as the redox potential indicates and if hydrochloric acid is used chlorine is given off.

$$VO_2^+ + 2H^+ + e^- \longrightarrow VO^{2+} + H_2O \quad E^{\ominus} = +1{\cdot}0\ V$$

$$VO^{2+} + 2H^+ + e^- \longrightarrow V^{III} + H_2O \quad E^{\ominus} = +0{\cdot}36\ V.$$

A further reduction to V^{III} is thermodynamically favourable, as the appropriate potential shows, but the product, the blue $[V^{III}(OH_2)_6]^{3+}$ aquo-ion, is stable only in the absence of air, as it reacts readily with oxygen to give the vanadyl cation VO^{2+} once more.

The redox chemistry of vanadium is neatly typified by the anaerobic reduction of vanadium(V) with zinc in acid solution. A series of well-defined colour changes occur, indicating the formation of vanadium(IV), vanadium(III), and finally vanadium(II). (Compounds of vanadium in lower valence states +1, 0, −1 are known but are completely unstable in water.) If air is then admitted the deep violet $[V^{II}(OH_2)_6]^{2+}$ is rapidly oxidized, first to vanadium(III) and then more slowly to give the VO^{2+} cation as before.

From these sequences it becomes clear just why the VO^{2+} ion dominates so much of the aqueous chemistry of vanadium. It persists throughout a wide series of reactions and gives rise to many five-coordinate, square-pyramidal complexes of the stoichiometry $(VOL_4)^{n+}$. However, this type of structure is by no means obligatory and many distorted octahedral ions such as $[VO(OH_2)_5]^{2+}$ are also known.

The biological role of vanadium.

Although the chemistry of vanadium is now quite well understood its biochemical role still remains uncertain. It has been known for about fifty years that some invertebrates accumulate vanadium. The ascidian worm *Phallusia mammillata* has been studied quite extensively and analysis of its blood has revealed vanadium concentration of up to 1900 p.p.m., which represents a concentration factor of over one million with respect to the sea-water in which it lives. The related organism *Ascidia nigra* has been reported to be an even more spectacular accumulator with concentrations of up to 1·45 % (14 500 p.p.m.) of vanadium in its blood cells.

Absorption of vanadium by these ascidian worms occurs principally through the branchial net (the ascidian equivalent of lungs) which is exposed to the sea-water. In this water the vanadium may exist either as oxovanadium(IV) or as various polyvanadate(V) ions. Tracer studies with ^{48}V(a β^+ emitter; half-life 16 d) have shown that it is only when it is in the higher oxidation state that vanadium is taken up by these organisms. Accumulation appears to be an active process since it is inhibited by the cardiac glycoside oubain, and shows a significant temperature dependence. It has been suggested that the polyvanadate ion is accumulated in mistake for phosphate and certainly the two processes are closely related, as excess phosphate or arsenate, but not chromate or molybdate, are also inhibitory. The accumulated vanadium is not found free in the blood, but instead is confined to vanadophores, which are distinct regions of the vanadium-carrying cells, the vanadocytes. Within these cells, the vanadium is believed to exist as haemo-vanadin, a non-haem vanadium-protein complex. The oxidation state of the vanadium within the vanadocytes remains uncertain, but as these cells are most unusual – containing as they do a considerable amount of sulphuric acid and having pH$\sim$0 – it is most unlikely that any vanadium(V) species persist, although vanadium(IV), as VO^{2+}, should be stable. It is impossible to rule out the possibility of further reduction to vanadium(III), and it has been claimed that a vanadium(IV)/vanadium(III) couple, which might be implicated in cellular reactions, has been detected in *Ascidia aspersa*.

Until recently, it was believed that vanadium played some part in the oxygen-transport cycle of the ascidians and other invertebrates, even though no oxygen-binding properties could be convincingly demonstrated *in vitro*. Now, however, this theory generally receives less credence, though definitive evidence is still lacking. Apart from the suggestion that the presence of vanadium is merely accidental (cf. PO_4^{3-}) it has also been postulated that it is an evolutionary residue. It is difficult to know just what weight to apply to these arguments, but it is interesting that oxovanadium(IV) porphyrins can be isolated from Venezuelan petroleum and that the ash from Peruvian asphalt contains up to thirty per cent of the element; though it is always possible that the vanadium was introduced not directly from fossil invertebrates, but by a geological process at some later date.

The absorption of vanadium by higher animals is poor, typically 0·1 – 1·0 per cent the ingested dose, which is fortunate as it is quite toxic. Nevertheless, a small amount has been consistently detected in most animals and also in man himself. Here the evidence for a functional role is much more convincing. Severe vanadium deficiency in laboratory animals leads to malformation of the bones and it appears that, under normal conditions, vanadium contributes to the mineralization process, perhaps by replacing phosphorus at specific sites. There is also some (albeit disputed) evidence that vanadium induces a resistance to dental caries. Vanadium also reduces the amount of cholesterol in the blood-stream by inhibiting its synthesis. The exact mode of this inhibition is still uncertain but it is known to occur fairly early in the biosynthetic pathway since vanadium prevents the incorporation of acetate or mevalonate, but not the cyclization of squalene.

Chromium ($3d^5 4s^1$)

Complexes of this element in all oxidation states from +6 to −2 are known, but only the +6, +3, and +2 states need be considered here since these are the only ones which are likely to be important for *in vivo* conditions. In the highest oxidation state, the best-known compounds are the chromate and dichromate anions which are linked by a pH-dependent equilibrium:

$$H_2Cr_2O_7 \underset{H^+}{\overset{OH^-}{\rightleftharpoons}} Cr_2O_7^{2-} + H_2O \underset{H^+}{\overset{OH^-}{\rightleftharpoons}} 2CrO_4^{2-}.$$

Acid solutions containing the dichromate anion are powerful oxidants and react with most organic materials, whereas in alkaline solution the chromate anion is only very weakly oxidizing. The change in redox potential is best illustrated by the following cycle:

$$2Cr^{III} + 6OH^- \rightarrow 2[Cr^{III}(OH)_3] \xrightarrow[E^{\ominus} = -0{\cdot}12V]{3H_2O_2 + 4OH^-} 2[Cr^{VI}O_4]^{2-}$$

$$\uparrow \qquad\qquad\qquad\qquad\qquad\qquad\qquad\qquad \downarrow\uparrow$$

$$2Cr^{III} \xleftarrow[E^{\ominus} = +1{\cdot}33V]{8H^+ + 3H_2O_2} [Cr_2^{VI}O_7]^{2-}$$

Hydrogen peroxide readily oxidizes chromium(III) to chromium(VI) in alkaline solution, but under acid conditions it is itself oxidized to oxygen with the concomitant reduction of chromium(VI) back to chromium(III).

The chemistry of chromium (III) has been extensively studied and a large number of regular mononuclear octahedral complexes are known. They are all paramagnetic with a magnetic moment approaching the spin-only value of $3{\cdot}88\ \mu_B$ expected of the $t_{2g}^3 e_g^0$ configuration which they must adopt. The considerable CFSE which results from this configuration makes the complexes of chromium(III) kinetically inert so that the rate of exchange of a water molecule from the first coordination sphere of $[Cr^{III}(OH_2)_6]^{3+}$ is about 2×10^{15} times slower than the corresponding exchange reactions for the corresponding chromium (II) complex.

The Cr^{3+} ion is a typical 'hard' Lewis acid and binds readily to both oxygen and nitrogen donor ligands. However, since the nitrogen donors produce a stronger ligand field and hence a greater CFSE, there is a tendency for chromium(III) to prefer such ligands. For example the total stability constant for $[Cr^{III}(NH_3)_6]^{3+}$ in aqueous solution has been determined as 2×10^{13}, which means that this hexammine complex is more stable than the hexaquo species by some 126 kJ mole^{-1} and the difference in CFSE

$$-\frac{6}{5}\Delta(NH_3) + \frac{6}{5}\Delta(H_2O) \cong 84 \text{ kJ mole}^{-1}$$

accounts for much of this difference.

Complexes of chromium(II) are well known and, with the exception of $[Cr^{II}(CN)_6]^{4-}$, which is low-spin, most have magnetic moments of about 4·9 μ_B, which is close to the value predicted for a high-spin $t_{2g}^3 e_g^1$ configuration. Because of this they are subject to a Jahn-Teller distortion (see p. 83) and are normally pseudo-octahedral, with four short and two long bonds. The majority of the complexes are stable only in the solid state or in the absence of air since chromium(II) is readily oxidized to chromium(III) as the appropriate redox potential indicates:

$$Cr^{III}_{(aq)} + e^- \rightarrow Cr^{II}_{(aq)} \qquad E = -0{\cdot}41 \text{ V}.$$

It is unlikely, therefore, that macro-quantities of chromium(II) complexes will be encountered under biological conditions. However, there are sufficient biological reducing agents of suitable redox potential to suggest that trace quantities of chromium(II) might be found and though these would be difficult, if not impossible, to detect, the possibility of their existence cannot be ignored, since even very small amounts of chromium(II) complexes can considerably increase the rate of ligand substitution reactions of the chromium(III) species. The mechanism of this catalytic effect has been shown to involve a cyclic reduction-oxidation process. The addition of ^{51}Cr (electron capture; half-life 28 d), as a tracer, as well as a chromium(II) complex, to a solution of a chromium(III) complex, is followed by the fairly rapid distribution of the radioactive label throughout the species in solution caused by the following equilibrium:

$$^{51}Cr^{II*} + Cr^{III} \rightarrow {}^{51}Cr^{III*} + Cr^{II}$$

The mechanism of such reactions has been extensively studied and has been shown to occur primarily via a process of atom transfer.

$$^{51}Cr^{II*} + X - Cr^{III} \rightarrow {}^{51}Cr^{III*} - X + Cr^{II}$$

Such reactions clearly account for the catalytic effect of traces of chromium(II). Other reducing agents exert similar effects and it is common practice to 'labilize' the reactions of chromium(III) by adding pieces of zinc metal or active charcoal to the reaction mixture.

The terrestrial distribution of chromium

Chromium, like most of the transition elements, with the obvious exception of iron, is a minor component of the earth's crust. Typical levels range from 100 p.p.m. in igneous rocks, to 10 p.p.m. or less in limestones. The chromium levels in soils derived from these rocks are more variable, and may be particularly high in highly weathered soils derived from basalt or serpentine; such soils may have specially adapted flora. The ready precipitation of chromium, as complex chromium(III) hydrated oxides or hydroxides, serves to keep the concentration of this element to vanishingly small levels in surface and ground water, with values rarely exceeding 200 p.p.b.

The biological role of chromium

It is only relatively recently that chromium was recognized as an essential element and so far the requirement has been conclusively demonstrated only in higher animals. It forms an essential part of the glucose tolerance factor GTF, which, together with insulin, is responsible for controlling the clearance of glucose from the blood-stream. Chromium has also been implicated in the control of cholesterol and lipid biosynthesis. In one study, a chromium supplement was shown to reduce the level of cholesterol in the blood-stream by 14 per cent. Chromium also affects amino-acid and nucleic-acid synthesis, and chromium-deficient animals often show genetic disorders.

The difficulty of studying the biological role of chromium is underlined by the fact that it was found to be effective at doses as low as 0·1 μg per 100 g of body weight, yet the lethal dose of chromium(III) has been estimated as only 1 mg per 100 g body weight. However, by careful control of the total environment it has been possible to show that rats fed on a low-chromium diet, such as Torula-yeast, have severely impaired glucose tolerance, as measured by the rate of clearance of an intravenously supplied glucose load, and actually excreted glucose in the urine. The condition is rapidly improved by supplying chromium, either as an extract from brewers' yeast, or as inorganic salts.

The mode of action of this supplement is still not clear, but in view of the synergistic effect between chromium and insulin it has been suggested that the metal could possibly stabilize the insulin molecule; either directly, by complex-formation, or indirectly by inhibiting the enzyme insulinase. Alternatively, it has been postulated that chromium could increase the binding of insulin to tissue and act as a cofactor for the membrane transfer of insulin.

A number of studies have indicated that the chromium in the GTF from brewers' yeast is chromium(III), but the actual nature of the 'complex' is still not clear. What seems to be certain is that it is not a regular octahedral species, since experiments with ^{51}Cr have shown that the chromium from the supplement exchanges extremely rapidly with that in the body pool. It is pertinent that the biological activity of simple complexes is restricted to those which undergo substitution reactions most readily and the highly stable, kinetically inert complexes such as $[Cr^{III}(en)_3]^{3+}$ have little effect. On the basis of known importance of Cr^{III}/Cr^{II} couple it is possible to suggest that the GTF contains chromium(III) in an environment which is conducive to the formation of chromium(II), which will 'labilize' substitution reactions. This idea has not yet been tested.

It is also likely that chromium plays a second independent role, since it has been shown to activate phosphoglucomutase, the enzyme which catalyses the conversion of glucose-l-phosphate to glucose-6-phosphate as the first step in the glycolytic pathway. This enzyme normally requires Mg^{2+} and another metal for full activity. Chromium is both the most efficient and the only one to sustain any activity in the absence of Mg^{2+}.

The efficiency of absorption of orally ingested chromium is low, generally between 1 per cent and 5 per cent, and it has been reported that compounds of chromium(VI) are absorbed more easily than those of chromium(III). It seems likely, though, that the difference is related more to the overall charge of the complex than any intrinsic difference in the reactivity of the metal itself, since the anionic complex $[Cr^{III}(C_2O_4)_3]^{3-}$ is absorbed more readily than the cation $[Cr^{III}(en)_3]^{3+}$. Simple salts of chromium, e.g. $CrCl_3$, are utilized only poorly, if at all, since the $[Cr^{III}(OH_2)_6]^{3+}$ ion which they form in solution is stable only in the acid of the stomach, and in the alkaline condition of the intestine it eventually forms highly insoluble polymeric hydroxides, which are metabolically inert. A similar reaction has been shown to occur during canning when, under the influence of heat, the normally stable protein-bound Cr^{III} undergoes olation,† so that tinned foods are a poor source of chromium. It is perhaps to prevent such reactions, as well as for other reasons, that chromium appears in the blood-stream bound to the protein transferrin, which is principally responsible for the translocation of iron.

Chromium is excreted mainly via the urine, in a process controlled by tubular reabsorption; this accounts for the high level of chromium in the kidneys. The level of chromium in the tissues shows a gradual, but so far unexplained decrease with age and it has been suggested that in view of the role of chromium in the GTF this might be one cause of maturity-onset diabetes. In support of this view, chromium has been shown to be effective in treating some, though by no means all, cases of this condition.

† Precipitation of salts of oleic acid, which is a major component of natural fats.

Manganese ($3d^5$ $4s^2$)

Manganese is known in no less than eleven oxidation states, ranging from +7 to -3, which is more than any other element; yet the aqueous chemistry of this element can be understood very easily since ten of these states are unstable with respect to the eleventh, manganese(II). (The lower oxidation states, -3 to +1, are found only in complexes with π-bonding ligands, such as carbon monoxide, and as these complexes decompose rapidly in water they need not be considered further.) The important compounds of the higher valence states +4 to +7 all have oxo-ligands and are oxidizing agents, the best known being the permanganate anion $[Mn^{VII}O_4]^-$, which formally contains the metal in the group oxidation state +7.

Solutions of the permanganate ion are intrinsically unstable and decompose slowly to form manganese(IV)oxide. The same product is also formed when permanganate reacts with reducing agents in neutral or alkaline solutions. Manganese(IV)oxide is itself a powerful oxidant, though it may appear unreactive by virtue of its insolubility in water. However, it reacts readily in acid solutions to give the expected Mn^{2+} ion.

The cyanide ion is one of the few ligands capable of producing spin-pairing in manganese(II); in other complexes the +2 ion nearly always adopts the high spin $t_{2g}^3e_g^2$ configuration. This means that there is no CFSE to be considered which simplifies considerably any discussion of the chemistry. Redox reactions excepted, manganese(II) can be treated as a simple cation, rather like an alkaline earth. The ionic radius of Mn^{2+}, 0·090 nm, is midway between that of Mg^{2+} and Ca^{2+}, so that the properties of Mn^{2+} can be predicted by extrapolating from these other two ions. For example, the rate of exchange of coordinated water from $[Mn^{II}(OH_2)_6]^{2+}$ is greater than that for Mg^{2+} but less than that for Ca^{2+} and, as noted before, Mn^{2+}can replace both calcium and magnesium *in vivo*.

The complexes of Mn^{II} are generally weakly coloured – usually pale pink. A closer examination of the spectrum shows that it consists of a large number of very narrow bands all with small extinction coefficients, the reason being that any movement of a d-electron in the $t_{2g}^3e_g^2$ configuration necessarily changes the spin multiplicity (the number of unpaired spins) and such changes are formally forbidden.

Compounds of Mn^{III} have received relatively little attention until recently, since the aquo-ion $[Mn^{III}(OH_2)_6]^{3+}$ is unstable and rapidly disproportionates to Mn^{2+} and MnO_2. The reaction can be slowed down by the addition of excess Mn^{2+} ion or by increasing the acid strength of the medium but, even so, manganese(III) remains a powerful oxidant. A few complexes, such as the acetylacetonate, $[Mn^{III}(Acac)_3]$ are known and judging by their magnetic moments, they contain the metal in the high-spin $t_{2g}^3e_g^1$, state. These complexes have an apparently regular octahedral geometry despite the supposed operation of the Jahn-Teller distortion for ions of this configuration. The low-spin configuration is known only in the cyanide complex $[Mn^{III}(CN)_6]^{3-}$.

The terrestrial distribution of manganese.

The level of manganese in the earth's crust is widely varied, being highest in calcareous sediments ($\gg$ 1000 p.p.m.) and least in highly weathered siliceous sediments such as sandstone ($<$ 50 p.p.m.). However, levels of manganese in rocks and soil are not directly indicative of its availability. The Mn^{2+} ion may be a major exchangeable cation in some very acid soils, but even if present in alkaline soil, it may not be readily available, since under these conditions it will be in the form of MnO_2, formed from manganese(II) hydroxide by bacterial oxidation. Interestingly though, it has been reported that the reverse process can be brought about by other soil bacteria which use the manganese(IV) oxide as a source of oxygen. The distribution of manganese is further modified by the associated levels of iron, since the geochemistries of these two elements interact quite strongly (see p. 102).

The biological role of manganese

Manganese appears to be indispensable to all organisms, both plant and animal. Manganese deficiency is well known in agriculture with the symptoms including chlorosis, since manganese is involved in the photosynthetic process. A simple quantosone, (the smallest morphological unit of the photosynthetic apparatus) has a relative molecular mass of about 2×10^6 and is thought to contain at least two manganese ions. It has been suggested that these function by cycling between the Mn^{III} and Mn^{II} states, resulting in the oxidation of water and the formation of molecular oxygen, in the final stage of the photosynthetic process. Certainly, the relevant oxidation potentials indicate that this reaction is thermodynamically quite feasible.

In regions where the soil manganese is low or not readily available, animals too tend to suffer from deficiency diseases, and common symptoms include malformation of the bones, infertility, and ataxia (loss of muscular coordination). The bones may develop incorrectly, being shorter and sometimes twisted, and this is ascribed to a failure in the development of the organic matrix. Whilst the condition is irreversible, the ataxia and infertility will respond to therapy with manganese salts. However, the detailed cause of these latter conditions is still uncertain.

Human beings also require manganese and have a body pool of between 12 mg and 20 mg. Studies with ^{54}Mn (electron capture; half-life 291 d) have shown that about 5 per cent of an orally ingested dose is absorbed, almost irrespective of the dose, but that the tissue levels remain unaltered as there is an extremely efficient homeostatic mechanism, operating principally by excretion of the unwanted ions. Further studies show that the manganese in the body is labile and that translocation occurs by means of transmanganin, a protein which forms a specific complex with tripositive manganese, though why this oxidation state is preferred and how the manganese reaches it remains a mystery.

The highest concentrations of the element are found in the liver, kidney, and the pancreas of human beings, where it is believed that the Mn^{2+} ion found in the mitochondria of these cells is possibly a cofactor for the respiratory enzymes.

In vitro, it has been shown that many enzymes which have only a loose requirement for metal ions are activated by manganese, often in competition with magnesium. Included in this category are many peptidases, phosphatases, and polymerases and specific enzymes such as galactransferase. Some enzymes contain more tightly bound Mn^{2+}, and amongst these pyruvate carboxylase has been particularly well studied. This enzyme, which also requires biotin and Mg^{2+} ions, catalyses the formation of oxaloacetate from carbon dioxide and pyruvate. The enzyme glutamine synthetase also contains Mn^{2+} ions. This enzyme is an oligomer consisting of 12 sub-units arranged in two hexagonal layers; the manganese is essential for the structural integrity of the assembled enzyme.

Iron ($3d^6\ 4s^2$)

The chemistry of iron in aqueous solution is relatively simple, and only the +2 and +3 oxidation states need be considered in any detail. When simple iron(II) salts with non-complexing anions are dissolved in water, they give rise to the pale green, very weakly acid aquo-ion $[Fe^{II}(OH_2)_6]^{2+}$, but similar salts of iron(III) are extensively dissociated as the following equilibria show:

$$[Fe^{II}(OH_2)_6]^{2+} \rightarrow [Fe^{II}(OH_2)_5\ (OH)]^{+} \qquad K_a = 10^{-9.5}$$

$$[Fe^{III}(OH_2)_6]^{3+} \rightarrow [Fe^{III}(OH_2)_5\ (OH)]^{2+} \qquad K_a = 10^{-3.0}.$$

The iron(III) salts are in fact stronger acids than ethanoic (acetic) acid and the pale purple $[Fe^{III}(OH_2)_6]^{3+}$ cation can exist only in the presence of added acid at $pH \leqslant 1$. Although the higher charge on the Fe^{3+} ion clearly accounts for the difference between the two iron salts, it is especially significant that iron(III) salts are also stronger acids than the corresponding aluminium(III) salts. The difference has its origin in the higher *effective* charge of the Fe^{3+} ion due to the poor shielding by the outer d-electrons; such discrimination between transition and main-group metals undoubtedly influences the selection of individual metal ions for particular biochemical functions. The iron(III) salts also undergo further hydrolysis and, in addition, show a distinct tendency to form polynuclear aggregates in which the iron atoms are linked by hydroxo-bridges. If further base (OH^-) is added these oligomers rapidly polymerize and form a colloidal gel of 'ferric hydroxide'.

$$[Fe^{III}(OH_2)_6]^{3+} + [Fe^{III}(OH_2)_6]^{3+}$$

$$\downarrow OH^-$$

$$[(H_2O)_4Fe^{III}(\mu\text{-}OH)_2Fe^{III}(OH_2)_4]^{4+}$$

$$\downarrow OH^-$$

'ferric hydroxide' gel

In the presence of complexing agents both Fe^{2+} and Fe^{3+} ions readily form stable complexes, mostly with an octahedral geometry, but whereas the Fe^{2+} ion binds strongly to nitrogenous ligands, Fe^{3+} has a very low affinity for such donors, unless (like *o*-phenanthroline (*o*-phen) or 2, 2-bipyridyl (bipy)) they are capable of inducing spin-pairing, so stabilizing the complex by the consequent gain in CFSE. Generally iron(III) has a greater affinity for oxygen-donor ligands such as phosphates, and polyalcohols such as sugars.

Unlike some of the earlier transition-metal ions, e.g. Cr^{3+}, which by virtue of their d-electron configuration are confined to one spin state, both iron(II) and iron(III) complexes are found with high- or low-spin configurations. The Fe^{2+} ion undergoes spin-pairing only under the influence of relatively strong ligand fields and, as expected, the change is characterized by a reduction in the magnetic moment from the typical high-spin value of 5·0 – 5·2 μ_B to zero. (This is higher than the predicted 'spin-only' moment 4·9 μ_B, due to a small orbital contribution.) More important, the change to a $t_{2g}^6e_g^0$ configuration substantially reduces the rate of ligand substitution reactions, and the complex becomes kinetically inert, as described previously. A most ingenious application of this result is the use of alkaline iron(II) tartrate as an antidote for cyanide poisoning. This treatment, which has proved to be effective (when applie rapidly enough!) relies on the formation of the inert $[Fe^{II}(CN)_6]^{4-}$ ion to 'mop up' the cyanide and thus render it harmless. In contrast, Fe^{3+} salts are ineffective and $[Fe^{III}(CN)_6]^{3-}$ is itself highly toxic as complexes containing the metal in this oxidation state are labile.

The spin-pairing energy of the d-electrons in the Fe^{3+} ions (358 kJ $mole^{-1}$) is nearly twice that of Fe^{2+} (270 kJ $mole^{-1}$), so that the only ligands capable of transforming iron(III) to a low-spin state are those which can produce extremely large orbital splittings (*o*-phen, bipy, CN^-). Consequently some ligands are capable of producing spin-paired iron(II) but leave iron(III) complexes in a high-spin configuration. Certain other ligands, notably the dithiocarbamates

$$R_2N{-}C\begin{matrix}{\diagup S^-}\\{\diagdown\!\!\diagdown S}\end{matrix}$$

produce orbital splittings such that the CFSE and the spin-pairing are finely balanced. Thus, depending on the experimental conditions, both high- and low-spin states are readily accessible within the same complex; the iron(III) complexes of such ligands have magnetic moments which are a complicated function of temperature. In most cases, however, the cross-over from a high-spin to a low-spin state is quite distinct, and is accompanied by a reduction in magnetic moment from 5·9 μ_B to ~ 2·3 μ_B.

The spectroscopic properties of iron complexes are less amenable to simple interpretation. The spectra of high-spin Fe^{2+} species show considerable broadening, or even a resolvable splitting, due to the Jahn-Teller distortion of the excited $t_{2g}^3e_g^3$ state. High-spin iron(III), which is isoelectronic with manganese(II), shows only very weak d-d transitions which are often obscured by charge-transfer bands. This is also the case with many low-spin iron(II) and iron(III) complexes.

The reaction of metallic iron with a non-oxidizing acid produces iron(II) salts and not those of iron(III), as inspection of the relevant oxidation potentials given in Table 7 would suggest. Conversion to the higher oxidation state may then be brought about by the use of an oxidizing agent. As expected, the strength of oxidant necessary to induce the change depends critically on the immediate environment of the Fe^{2+} ion. The effect of complex formation, and, in particular, the high-spin to low-spin cross-over predicted by the crystal field theory, is particularly well documented and as will be seen has particular biochemical significance, by altering the value of the redox potential for the Fe^{III}/Fe^{II} couple.

Fig.18 shows how the CFSE is at least partially responsible for the observed changes in redox potential shown in Table 7. Three cases can be distinguished, depending on the spin state of the metal ions. In the first, high-spin iron(II) gives high-spin iron(III), with a consequent loss in CFSE, of $(2/5)\Delta(Fe^{2+})$. Naturally, therefore, the larger $\Delta(Fe^{2+})$ the less favourable the reaction and the larger is $E^{\ominus}$. A sufficient increase in Δ may cause spin-pairing of the +2 ion, though not for the +3 state. This increases the stability of the lower oxidation state considerably, since oxidation results in a transformation from t_{2g}^6 to $t_{2g}^3e_g^2$ configuration which, in turn, results in a loss of CFSE of no less than $(12/5)\,\Delta\,(Fe^{2+})$. Finally, the value of Δ may increase beyond the point

TABLE 7

Some redox potentials for the Fe^{III}/Fe^{II} couple showing the variation caused by changing ligands

	Oxidized species	Reduced species	E (V)
†	$[Fe^{III}(o\text{-phen})_3]^{3+}$	$[Fe^{II}(o\text{-phen})_3]^{2+}$	+ 1·10
	$[FE^{III}(OH_2)_6]^{3+}$	$[Fe^{II}(OH_2)_6]^{2+}$	+ 0·77
	$[Fe^{III}(CN)_6]^{3-}$	$[Fe^{III}(CN)_6]^{4-}$	+ 0·36
	Cytochromes a (Fe^{3+})	Cytochromes a (Fe^{2+})	+ 0·29
	Cytochromes c (Fe^{3+})	Cytochromes c (Fe^{2+})	+ 0·26
	Haemiglobin	Haemoglobin	+ 0·17
	Cytochromes b (Fe^{3+})	Cytochromes b (Fe^{2+})	+ 0·04
	Myoglobin	Myoglobin	+ 0·00
	$[Fe^{III}EDTA]^{-}$	$[Fe^{III}EDTA]^{2-}$	- 0·12
‡	$[Fe^{III}(oxine)_3]$	$[Fe^{II}(oxine)_3]^{-}$	- 0·20

† *o*-penanthroline.
‡ oxine = 8-hydroxyquinoline.

at which spin-pairing occurs for the Fe^{3+} ion. This means that there will now be an appreciable stabilization of the higher oxidation state, and therefore the oxidation potential should be reduced.

Of course, this description has taken no account of changes in spin-pairing or exchange energies, which may reduce the predicted effects, but the relative order of changes is expected to remain unchanged by these factors. However, other factors must also be considered; one of the more important is the stabilization of the iron(III) state by anionic ligands. This is an entropy effect in which the binding of anionic ligands reduces the interaction of the metal ion with water and, therefore, reduces the hydration of the complex as a whole, consequently increasing the entropy of the system. This is one of the reasons why $[Fe^{III}(EDTA)]^{-}$ (EDTA = ethylenediaminetetracetic acid) is actually stable with respect to the reduced complex species $[Fe^{II}(EDTA)]^{2-}$, yet $[Fe^{III}(OH_2)_6]^{3+}$ is unstable towards reduction to $[Fe^{II}(OH_2)_6]^{2+}$ despite the fact that the overall ligand fields produced by $EDTA^{-4}$ and $6H_2O$ differ very little. Deviations from these general

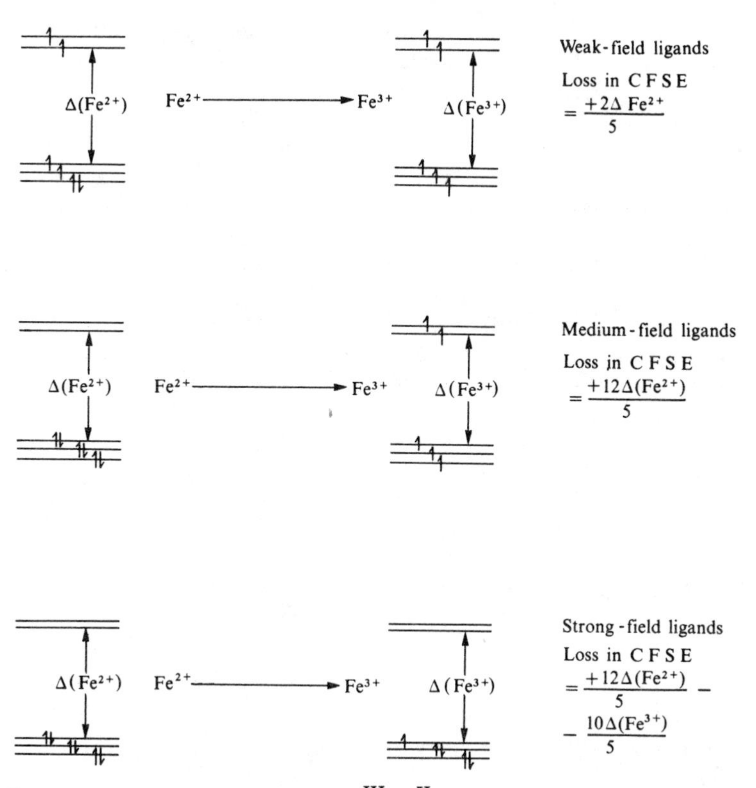

FIG.18. The redox potential of the Fe^{III}/Fe^{II} couple is substantially modified by the nature of the ligands.

rules do occur, particularly if, in either oxidation state of the metal, the complexes have less regular stereochemistry.

The terrestrial distribution of iron

Although it seems to be accepted that the bulk of terrestrial iron is concentrated in the earth's core, it is still one of the most abundant metallic elements of the lithosphere, with typical concentrations of over 50 000 p.p.m. in primary igneous rocks. The geochemistry of iron is intimately connected with its redox chemistry and can largely be summarized in the following statement: acid anaerobic conditions favour the formation of soluble, mobile iron(II) salts, whereas alkaline aerobic conditions favour its immobilization in various modifications of iron(III) oxide.

The distribution and availability of iron are subject to considerable modification by the process of weathering. For example, iron is easily leached from sulphide ores via the oxidative formation of the soluble iron(II) sulphate and it is also removed quite readily from the silicate minerals as the hydrogen carbonate. Once in solution iron(II) salts are easily oxidized to iron(III) and under even the mildest alkaline conditions the iron in this state is re-precipitated as the oxide. The oxidation may occur directly through molecular oxygen or indirectly through a variety of other oxidizing agents. Particularly important in this respect are the higher oxides of manganese which act as oxygen-transfer agents. As a consequence the availability of iron depends considerably on the levels of manganese. This principle operates both in the soil and in the sea which, by virtue of its alkaline nature ($pH \geqslant 8$), contains little soluble iron ($< 0{\cdot}01$ p.p.m.). Instead much of the leached iron reaching the sea is co-precipitated with manganese in the form of nodules consisting of about one-third MnO_2 and one-fifth Fe_2O_3.

Despite the continuous leaching process, few soils are actually iron-deficient, the exceptions being those derived from limestone strata which have inherently low iron levels and those which are derived from acid peats, from which the iron has been almost completely leached by water logging. On the other hand many soils are known which are *effectively* iron-deficient by virtue of the unavailability of the iron.

The biological role of iron

Iron appears to be an essential element for all organisms, both plant and animal. In plants, iron deficiency is not uncommon, particularly on alkaline or high manganese soils as already indicated. Thus the failure of such calcifuge species as rhododendrons to grow well in calcareous soils is largely attributable to their inability to absorb sufficient iron under these conditions. The problem can be overcome by supplying iron in the form of a stable complex such as $[Fe^{II}EDTA]^{2-}$, from which the iron is not readily precipitated. The symptoms of iron deficiency are variable but usually include chlorosis, though once again chlorosis does not necessarily indicate iron deficiency.

Despite the variable occurence of iron in plants, uncomplicated iron deficiency in grazing stock is almost unknown, but numerous other iron-deficiency states associated with imbalances in the metabolism of elements such as cobalt or copper are well documented. In man it has long been apparent that anaemia (literally 'no blood') results from a dietary deficiency of iron and can be alleviated by administering an inorganic supplement. This has resulted in many long and detailed studies of all the facets of iron biochemistry.

In early studies on the treatment of anaemia, iron was administered orally in the form of simple salts, but it was soon found that only relatively massive doses were effective, since the majority was rapidly excreted. Subsequently, tracer studies with ^{59}Fe (a β^- emitter; half-life 44·3 d) have shown that only 5 – 10 per cent of dietary iron is absorbed and from this it has been estimated that, on average, man absorbs ~ 1 mg of iron per day from a typical diet containing about 15 mg of available iron. Since the average adult contains a total of over 4 g of iron in various metabolic fractions, it is clear that the human body must be able to utilize this element most economically. For example, if the average life of an erythrocyte is taken to be 120 d, this leads to the conclusion that the daily iron requirement of the blood-forming tissues alone is ~ 25 mg; although this is linked to changes in other iron pools, a turnover of ~ 20 mg per day is indicated, yet all this is kept in balance by the absorption of 1 mg daily. The principal pathways involved in the metabolism of iron in man are shown in Fig. 19.

The absorption of iron occurs primarily in the duodenum and to some extent this process depends on the chemical state of the iron. The actual mechanism of absorption is still a contentious subject, but there are

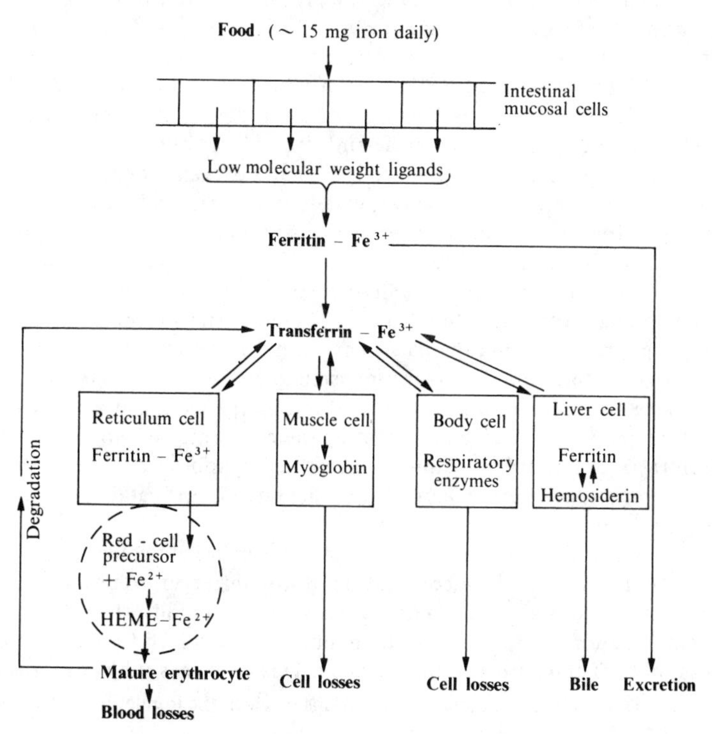

FIG.19. A simplified view of iron metabolism.

clearly both active and passive processes involved. The active absorption has been shown to occur via the mucosal cells of the intestine. These have a limited iron-storage capacity and if the ingested level is high they become saturated, with consequent inhibition of the absorption process. This is commonly referred to as the 'mucosal block theory'. The passive uptake of iron can be controlled by the chemical form in which the iron is presented. Passive absorption of simple salts is poor, which is hardly surprising in view of the difficulty with which cations are translocated across membranes, but by presenting the iron in the form of neutral or anionic complexes, uptake is dramatically increased. Interestingly, the absorbed iron is fairly mobile and, by using a $^{59}Fe^{3+}$ - fructose complex, it has been shown that the time taken for the iron to appear in the extremities is no longer than the time taken for the circulation of the blood.

As the iron passes into the blood from the stomach, the change from strongly acid to very mildly alkaline conditions (pH 7·4) means that any iron(II) is more readily oxidized to iron(III), and it is in this state that the metal is next found, bound to the protein transferrin (which has been mentioned previously as being concerned in the transport of chromium). It is currently believed that transferrin is not responsible for the actual absorption process, but functions chiefly in the translocation and utilization of the iron. This view is supported by studies of atransferrinaemia (the congenital absence of transferrin). Patients with this condition continue to absorb the iron which is found as excessive deposits in the liver and associated organs whilst, paradoxically, the victims exhibit all the symptoms of chronic iron-deficiency anaemia.

Each molecule of transferrin (m.w. = 90 000) is capable of binding two atoms of iron and the stability constant of the transferrin-Fe^{3+} complex has been estimated at 10^{26}. Therefore transferrin serves to scavenge free iron from the plasma and can even remove it from the otherwise stable complexes with phosphate or citrate ions. The effect of this is to keep the concentration of iron in the plasma at a concentration below 1 p.p.m, thus protecting the cells which do not require iron from an unwanted loading, which could possibly lead to siderosis.† More important, it also conserves iron liberated in the catabolism of erythrocytes.

Recent studies have shown that the transferrin-Fe^{3+} complex is directly involved in the incorporation of iron into erythroblasts (immature red cells). The complex was shown to bind to the cell membrane for a period of about a minute, during which time the iron was split off and absorbed into the cell. This process is subject to a

† A general term for a number of conditions in which the iron level is substantially raised.

highly sophisticated control mechanism. The apo-transferrin binds less strongly to the cell than the transferrin-Fe^{3+} complex and thus is displaced by a fresh carrier molecule. Not only this, but as the cell approaches maturity transferrin no longer binds to the cell at all, so preventing excessive intake of the metal.

Iron is stored in the liver, spleen, and bone marrow by means of the related proteins ferritin and haemosiderin. Ferritin, which is water soluble and may readily be crystallized, consists of a protein 'coat' surrounding a micelle of iron(III) hydroxyphosphate which may be 90·0 nm in diameter. The total iron content of ferritin isolated from horse spleen is ~ 20 per cent. The exact nature of the other iron-protein, haemosiderin, is still uncertain and although the protein moiety is very similar to apo-ferritin, the iron content, estimated at about 40 per cent, is significantly higher.

The transfer of iron between transferrin and ferritin is an active process requiring both ATP and ascorbic acid. The ATP provides the energy whilst the ascorbic acid acts as a reducing agent, yielding iron in the +2 state; it is in this form that the iron is first incorporated into ferritin. This seems rather peculiar, particularly in view of the kinetic inertness of iron(II) complexes; and it appears that much is yet to be learnt about the behaviour of metals in systems of macromolecular ligands.

Almost three-quarters of the iron in a human being is in the form of the oxygen-carrying pigment, haemoglobin. Treatment of haemoglobin with hydrochloric acid in acetone in the absence of oxygen yields two fractions, one a protein (globin), and the other the iron-containing unit, haem. Fig.20 shows part of the structure of the haemoglobin isolated from human blood.

Haem itself, which is formulated as a high-spin iron(II) complex in accordance with a wide variety of magnetic, spectroscopic, and chemical data, reacts readily with a variety of additional ligands. The addition of one axial substituent encourages the binding of a sixth ligand to give an octahedral haemochrome, a process which is to be expected, judging by the preferred geometry of simple iron(II) complexes. In the presence of oxygen or other oxidizing agents, haem and the various haemochromes are readily oxidized to haemin and haemichromes respectively. The oxidation potential of this Fe^{III}/Fe^{II} couple depends intimately on the nature of the axial ligands and, as predicted from simple crystal field theory, the stability of the +2 oxidation state increases with the increasing field strength of these ligands, with $E^{\ominus}$ rapidly becoming more positive as these ligands cause a cross-over from high- to low-spin for the iron(II) state.

The redox potential can also be subtly affected by variations in the substituents of the porphyrin ring. Such substituents may withdraw or donate electrons to the ring system, thus altering the electron density at the pyrrole nitrogen atoms and thereby influencing the d-orbital splitting.

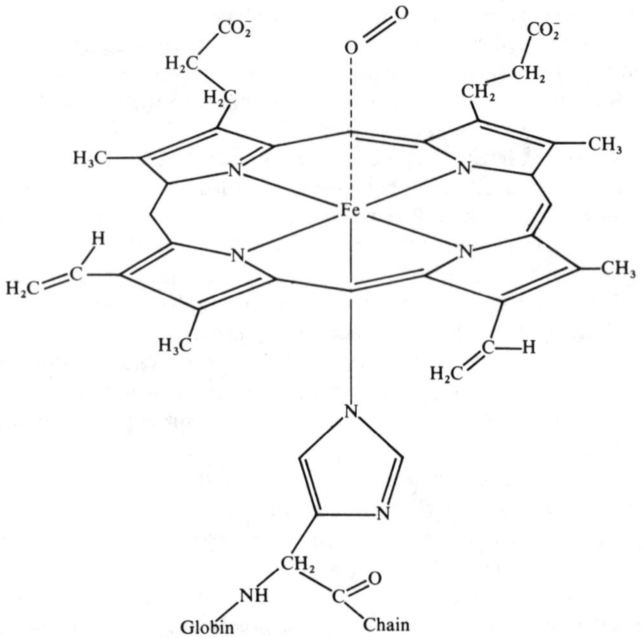

FIG.20. Oxyhaemoglobin: the oxygen-carrying form of human haemoglobin. The haemoglobin molecule is actually a tetramer, consisting of four haem (ferroprotoporphyrin IX) units and four protein (globin) chains. The haem prosthetic group is also found in human myoglobin and cytochromes of class B and also in enzymes such as catalase and peroxides. Compare this structure with that of chlorophyll (Fig.14) and Vitamin B_{12} (Fig.24).

This is believed to be highly significant in determining the exact redox potential of the cytochromes.

The active form of haemoglobin is a tetramer containing four haem units, each bound loosely by non-covalent forces to a globin chain. The imidazole nitrogen from a histidine residue in each protein chain provides the fifth ligand for the iron and, in the absence of oxygen, the sixth position is probably occupied by a water molecule. Examination of the oxidation potential of deoxyhaemoglobin shows that it is actually more stable with the iron in the Fe^{3+} state and indeed haemoglobin, like the simple haem prosthetic group, is readily oxidized by a wide variety of one-electron oxidizing agents to haemiglobin (methaemoglobin) which is a typical high-spin iron(III) complex with a magnetic moment of 5·7 μ_B, corresponding to 5 unpaired electrons.

What is most remarkable, however, is that oxygen is normally unable to induce this change, despite being able to oxidize the free haem group quite readily. Instead, the oxygen molecule displaces the water molecule and binds to the sixth coordination position, thus bringing about a change from high-spin to low-spin iron(II), as shown by the fact that oxyhaemoglobin is diamagnetic. Clearly then, the very act of oxygen-binding must increase the ligand field quite enormously, perhaps to such a value that the oxidation potential of the Haem-Fe^{III}/Haem-Fe^{II} couple also increases and passes beyond the limit at which oxygen can bring about the oxidation reaction.

Even if this is not the case and the reaction is thermodynamically allowed, it may well prove to be kinetically forbidden and here the difficulty of placing a single extra electron onto the oxygen molecule is doubtless most important. Note that this problem does not arise in the case of the free haem group since the reaction may be written:

$$[\text{Haem-Fe}^{II}]^{2+} + O_2 \rightarrow [\text{Haem-Fe}^{II} \leftarrow O_2]^{2+} + [\text{Haem-Fe}^{II}]^{2+}$$

$$\downarrow$$

$$2[\text{Haem-Fe}^{III}]^{3+} + H_2O_2 \xleftarrow{2H^+} [\text{Haem-Fe}^{II} \leftarrow O_2 \rightarrow \text{Fe}^{II}\text{-haem}]^{4+}$$

Clearly this is impossible for haemoglobin, as the haem prosthetic groups are separated in the molecular structure.

All four haem units in the haemoglobin molecule are capable of binding

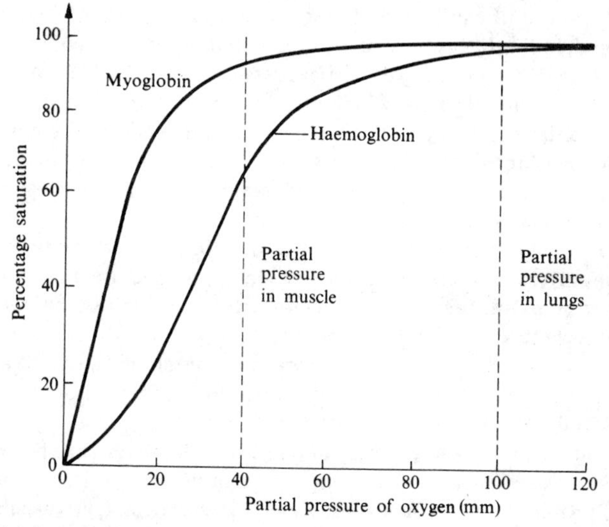

FIG.21. Oxygen-binding properties of haemoglobin and myoglobin. From these it can be seen that myoglobin is more readily saturated than haemoglobin.

an oxygen molecule, but it has long been established that they do not function independently and that the binding of each oxygen affects the binding of the subsequent one, so that the relative affinities are 1:4:24:9. This phenomenon, known as cooperativity, produces the typical sigmoid curve for O_2 binding shown in Fig.21.

The origin of the cooperativity lies in the change in spin state caused by binding a molecule of oxygen. Deoxyhaemoglobin, which contains iron in the high-spin ($t_{2g}^4 e_g^2$) configuration, has the iron atom some 0·07–0·08 nm above the plane of the porphyrin ring since it is just too large to fit the 'hole' in the macrocyclic ligand. On binding oxygen, an obligatory change to a low-spin $t_{2g}^6 e_g^0$ configuration occurs, which reduces the size of the iron sufficiently to allow it to slip neatly into place in the plane of the ring. This movement 'pulls' the attached histidine residue into a new position and thus alters the conformation of the whole globin chain, and it is this allosteric change which causes the observed cooperativity.

Haemoglobin is contained in the erythrocytes and serves to transport molecular oxygen from the lungs to the tissues. Here, the lower partial pressure of the oxygen brings about a reversal of the binding equilibrium and oxygen is released. However, rather than just leaving it to dissolve in the cellular fluids and then be free to diffuse away in all directions, it is rapidly bound to another haem-iron protein, myoglobin.

Myoglobin contains one haem group and has a molecular weight of 17 000, so that it can be loosely regarded as the monomeric equivalent of haemoglobin. It has only one oxygen binding site and consequently binds oxygen better at lower partial pressures (Fig.21). It is contained mainly in the muscle cells, which of course require most oxygen. It is typical of the efficiency of the evolutionary process that there is a very large myoglobin content in the tissues of whales and other diving mammals, which need to store considerable amounts of oxygen directly in the muscle tissue.

The oxygen-carrying pigments of some primitive vertebrates such as the lamprey eel also contain a single haem group and are closely related to myoglobin but, in contrast, some invertebrates such as the polychete and oligochete annilid worms have oxygen carriers with molecular weights of over 10^6. These *erythrocruorins* contain between 30 and 400 haem groups per molecule but, regardless of size, they all bind one oxygen molecule per haem.

Not all respiratory pigments contain haem, however, and the worm *Sipunculus* uses *haemerythrin* instead, which in place of the haem unit has 16 protein-bound iron atoms which, between them, are capable of binding eight molecules of oxygen.

The haem unit is obviously a most efficient evolutionary invention since it is not restricted to just the oxygen-carriers but is also found in

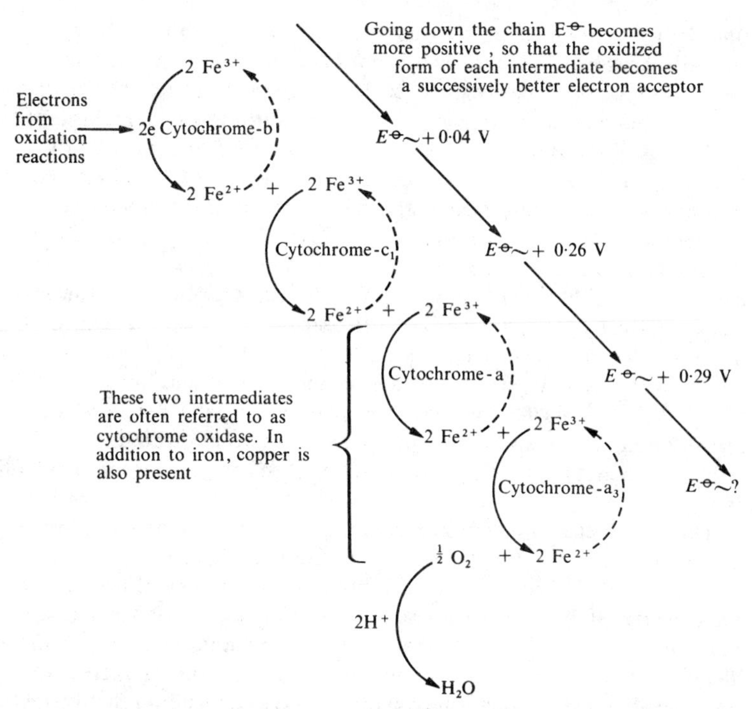

FIG.22. The electron-transport chain of the mitochondrion. The electrons produced by the oxidation of such substrates as pyruvate, isocitrate, or malate are passed through non-haem carriers such as FAD and co-enzyme Q to the cytochromes, and hence to the ultimate electron electron acceptor, the oxygen molecule.

other types of functional compound. A second category of haem-proteins are the hydroperoxidase enzymes. This class includes catalase, peroxidase, and cytochrome-c peroxidase, all of which are widely distributed in nature. They are normally isolated with the iron as high-spin Fe^{3+}, and their function is to decompose a peroxide, which they do without a formal change in valence.

The third class are the cytochromes, haem-protein molecules which act as electron carriers, linking the oxidation of substrate with the reduction of molecular oxygen in aerobic metabolism, Fig. 22. In contrast to the other haem compounds, the cytochromes function by shuttling between the iron(II) and iron(III) states and act without directly binding the substrate molecule.

The classification of cytochromes is complex since they differ from organism to organism and even within one particular species. Also iso-cytochromes, cytochromes which differ only in a few of the amino-acid units of the protein chain, are well known. Generally though the cytochromes can be divided into two major types. The first, cytochromes b_1, have low oxidation potentials, suitable for interaction with the substrate. The prosthetic group is protohaem and the reduced form contains iron as low-spin iron(II). Both the fifth and sixth ligands are provided by the protein so that cytochromes of this class are incapable of binding oxygen or other potentially interfering molecules such as carbon monoxide.

The second class, cytochromes c, are distributed ubiquitously from the simplest to the most complex organisms and because of this it has been suggested that these cytochromes were developed early in the evolutionary time scale and proved so efficient that they have remained unmodified ever since. The c cytochromes also use the protohaem prosthetic group, which is bound to the protein moiety by covalent linkages between the vinyl groups on the porphyrin ring and two cysteine residues. It is additionally bound by coordination of the iron by two more amino-acid residues which renders this cytochrome inert to oxygen, carbon monoxide, and cyanide anions.

The fourth class of electron carriers, the cytochrome oxidases, provide the link between the cytochromes c and the final electron acceptor, oxygen, but their structure is not fully known at present. However, it has been clearly established that these molecules contain both haem groups and copper(II) ions. They are capable of binding molecular oxygen and are therefore inhibited by carbon monoxide or cyanide ion. The effective reducing ability of cytochrome oxidase seems to depend on the interaction of the different metal ions within the molecule.

Although the importance of the haem group has long been established, it is only recently that the equal importance of non-haem iron proteins (NHIP), and the ferredoxins in particular, has been recognized. The name ferredoxin was first applied to the NHIP isolated from the anaerobic bacterium *Clostridium pasteurianum*, where it was shown to act as an electron carrier in the nitrogen-fixation process. *Cl. pasteurianum* does not have any photosynthetic ability and so the bacterial ferredoxin was not immediately connected with another NHIP electron carrier which had previously been isolated from spinach chloroplasts. This molecule was originally given the name 'haem-reducing factor' and was shown to be part of the photosynthetic apparatus. Eventually, however, the weight of evidence suggested that these two substances were similar, though not quite identical and they were given the general name *ferredoxin*.

This class of NHIP has now been isolated from non-photosynthetic anaerobes, photosynthetic bacteria, algae, and green plants. A study of the amino-acid sequences of the protein chains suggests that this was the order of evolutionary development, which concurs with the widely held belief that photosynthesis occurred first in the development of life and that this produced oxygen which then made possible the emergence of

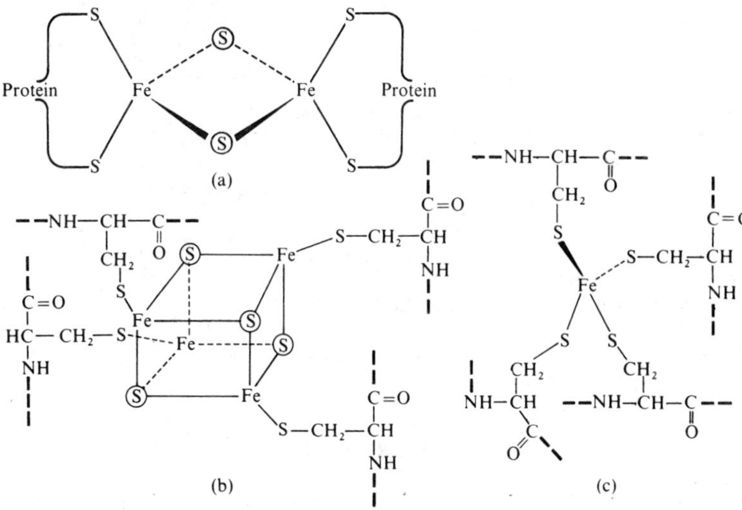

FIG.23. Possible structures for some non-haem iron–sulphur proteins involved in biological redox reactors:

(a) the $2Fe - 2S^{2-}$ unit of plant ferredoxins;
(b) the $4Fe - 4S^{2-}$ unit of bacterial ferredoxins (many bacterial ferredoxins contain two such units);
(c) the distorted tetrahedral structure of the NHIP, rubredoxin, isolated from some aerobic photosynthetic bacteria.

aerobic life-forms. The ferredoxins are relatively small protein molecules, having molecular weights in the range 6000–12 000, dependent on their source, and the number of iron atoms ranges similarly from 2 to 8. The most unusual feature of the ferredoxins is that they liberate hydrogen sulphide when treated with mineral acids. Fig.23 shows possible structures of this class of compound.

Clostridium pasteurianum contains a second NHIP, rubredoxin. This molecule has a molecular weight of about 6000 and contains one atom of iron, which is in a distorted tetrahedral environment, being bound by the sulphur atoms of four separate cysteine residues. It does not, however, contain any inorganic sulphide. There are considerable similarities between the protein chains of ferredoxin and rubredoxin and it has been suggested that they may have evolved from the same archetypal gene. More recently other related iron-sulphur proteins have been isolated and the properties of two of these, adrenodoxin and putidaredoxin, are listed in Table 8.

The oxidation potentials of all these compounds, except adrenodoxin, are negative, implying that the preferred state for the iron is as iron(III), which expressed in an alternative fashion means that in the iron(II) form

TABLE 8

The properties of some iron-sulphur-proteins: biological redox catalysts

Type	Source	Fe : S^{2-}	$E^{\ominus}$(V)
Bacterial ferrodoxin	*Azotobacter vinelandii*	8 : 8	-0·42
	Azobacter vinelandii	6 : 6	-0·32
	Bacillus polymyxa	4 : 4	-0·37
Plant ferredoxin	Spinach	2 : 2	-0·40
Putidaredoxin	*Pseudomona putida*	2 : 2	-0·23
Adrenodoxin	Mammalian	2 : 2	-0·34
Rubredoxin	Aerobic photosynthetic bacteria	4 : 0	-0·05
HPIP	*Chromatium*	4 : 4	+0·35

they readily lose an electron and so act as reducing agents, which is their principal biochemical function. Interestingly, though, even if they contain more than one iron atom they frequently appear able to accept only one electron. Putidaredoxin, for example, contains two atoms of iron and in the oxidized form of the compound both exist as low-spin iron(III), whilst in the reduced state the extra electron is shared between them. Green plant ferredoxins are also one-electron reductants, but the bacterial ferredoxins from *Clostridia* and *Chromatium* are two-electron acceptors.

One other structurally similar class of iron-sulphur proteins has been isolated. These high-potential iron proteins, HPIP, have positive potentials of up to +0·7 V. The best characterized comes from *Chromatium* and has a molecular weight of about 10 000. It contains four iron atoms, four atoms of labile sulphur, and four cysteine residues arranged in a tetrahedral array.

Cobalt ($3d^7\ 4s^2$)

Cobalt, like iron, is a typical electropositive metal and reacts with acids to form salts. In these compounds the metal adopts the dipositive state, and salts of cobalt(III), which usually have to be prepared indirectly, are rare. The ones which are known, such as CoF_3 and $Co_2(SO_4)_3$, all have anions derived from the electronegative elements such as oxygen and fluorine. Even these are stable only in the solid state, since in solution the metal is rapidly reduced to cobalt(II) whilst the water is oxidized to oxygen. The oxidant is actually the hexaquocobalt(III) ion which is an extremely powerful oxidizing agent as the relevant potential shows: $E^{\ominus} = +1{\cdot}8$V. Even at 0 °C this ion $[Co^{III}(OH_2)_6]^{3+}$ has a half-life of no more than one month. This decreases rapidly as the temperature increases and is only a few minutes at 25 °C.

The decreased stability of the higher oxidation state is in line with the trend already observed on moving across the transition series. It is reflected by the increase in ionization energies and is caused by the reduction in atomic radius concomitant with increased nuclear charge. Of course, the relative stability of the two oxidation states will be altered considerably by the presence of complexing agents, but even so most cobalt(III) complexes are thermodynamically unstable with respect to cobalt(II). Nevertheless a vast number of cobalt(III) complexes are known; the coordination chemistry of cobalt provides an excellent example of the difference between thermodynamic and kinetic stability.

Almost without exception these complexes are octahedral and adopt a low-spin, $t_{2g}^6e_g^0$ configuration. Consequently they benefit from a large CFSE, $(-12/5)\Delta$ and as a result are kinetically inert. This is particularly true if the ligand is polydentate and complexes such as $[Co^{III}\text{-EDTA}]^-$ can be boiled in both acid and base with little effect. It is a consequence of this kinetic inertness, rather than thermodynamic stability, that so many complexes of cobalt(III) have been prepared.

The large CFSE of the cobalt(III) must be compared with the much smaller value of $(-4/5)\Delta$ for the typical high-spin octahedral cobalt(II) species. (Note that Δ changes with the charge on the metal.) Clearly, therefore, the higher oxidation state should be stabilized by the presence of strong-field ligands, a prediction which is confirmed by experiment. (It is worth remembering that cobalt(III) is isoelectronic with iron(II) which is also stabilized by strong field ligands.)

Cobalt(II) forms a wide range of complexes which usually have octahedral or tetrahedral geometry, though a lesser number of square-planar or five-coordinate species are also known. In general, monodentate anionic ligands such as Cl^- produce complexes with tetrahedral geometry, whereas neutral ligands like H_2O normally form octahedral complexes. The two geometries are easily distinguished spectroscopically since the absence of a centre of symmetry in the tetrahedral complexes make the d-d transition less forbidden, so that the $[Co^{II}X_4]^{2-}$ chromophore usually has an extinction coefficient more than two orders of magnitude larger than an octahedral complex.

Almost without exception, the octahedral cobalt(II) complexes adopt a high-spin $t_{2g}^5e_g^2$ electronic configuration, but the magnetic moments, typically $4{\cdot}7 - 5{\cdot}2\,\mu_B$ are considerably higher than the spin-only value of $3{\cdot}87\,\mu_B$ and are not amenable to simple interpretation. The CFSE for such complexes, $(-4/5)\Delta$, is less than for cobalt(III), which is one factor which contributes to the kinetic lability of cobalt(II) complexes. The rate of substitution of water into the inner coordination sphere ($K = 5 \times 10^5\ s^{-1}$) is similar to that of magnesium ($K = 1 \times 10^5\ s^{-1}$) and the difference is consistent with their relative ionic radii ($Co^{2+} = 0{\cdot}074$ nm; $Mg^{2+} = 0{\cdot}065$ nm). Unexpectedly, then, the chemistry of Co^{2+} resembles that of Zn^{2+} rather than Mg^{2+}. This is peculiar, since despite the similarity in ionic radii ($Zn^{2+} = 0{\cdot}074$ nm) the rate of water exchange is considerably faster for Zn^{2+} ($K = 5 \times 10^7\ s^{-1}$) than for Co^{2+}. Nevertheless, Co^{2+} seems able to replace Zn^{2+} in enzymes most efficiently, at least *in vitro*, and the substitution often enhances the rate of the enzyme-catalysed reaction. It seems that the flexibility of the coordination of Co^{2+} is the controlling factor in this case. These similarities have made cobalt(II) a useful probe for several zinc-potentiated enzymes, as has already been mentioned.

The terrestrial distribution of cobalt

Cobalt, like the element which follows it, nickel, is geochemically very similar to divalent iron though with very much lower abundance. Typical values for levels of either element in the soil rarely exceed 50 p.p.m. and, more important, this means the extractable fraction is correspondingly small. The net result is that both can be considered to be bio-geochemicall[y] unavailable. In the case of cobalt this is unfortunate since in many areas o[f] the world it leads to a specific deficiency disease for grazing animals.

The biological role of cobalt

Cobalt was first shown to be an essential element for ruminants as the result of studies aimed at finding a cure for Coast disease (otherwise known as wasting sickness or bush sickness) which affects sheep in certain areas of Australia. This condition was endemic amongst sheep which grazed on apparently good pastures and, apart from moving them to alternative grazing, no cure was known. A primary symptom of this condition was anaemia and so iron therapy was applied but this met with only limited success. The treatment was found to be effective only when enormous doses of iron were administered, which themselves could prove fatal. On post-mortem examination the animals were found to be sufferin[g] from siderosis and had considerable deposits of iron in the liver and kidneys.

It was thought therefore that the partial success of the iron treatment was due to an impurity in the iron. Extensive tests confirmed this view and cobalt was eventually identified as the active component. Treatment of the pasture with cobalt salts or direct administration of cobalt to mature animals alleviated all the symptoms and the blood was rapidly restored to a normal condition. Strangely, however, when suckling lambs were dosed in a similar fashion a different response was noted:

erythropoiesis† was stimulated, and when the animals were given prolonged treatment they developed polycythaemia.‡ Similar results could be obtained by injecting cobalt into adult animals.

No explanation for these observations was satisfactorily advanced till over twenty years later when the anti-pernicious anaemia factor of raw liver was isolated in 1948. The efficiency of raw liver in the treatment of anaemia in man had long been known; but purification proved inordinately difficult partly because of the small amount of the factor present in liver and also because it was necessary to use anaemic patients to assay for the active principle. Eventually, starting from ton quantities of liver,

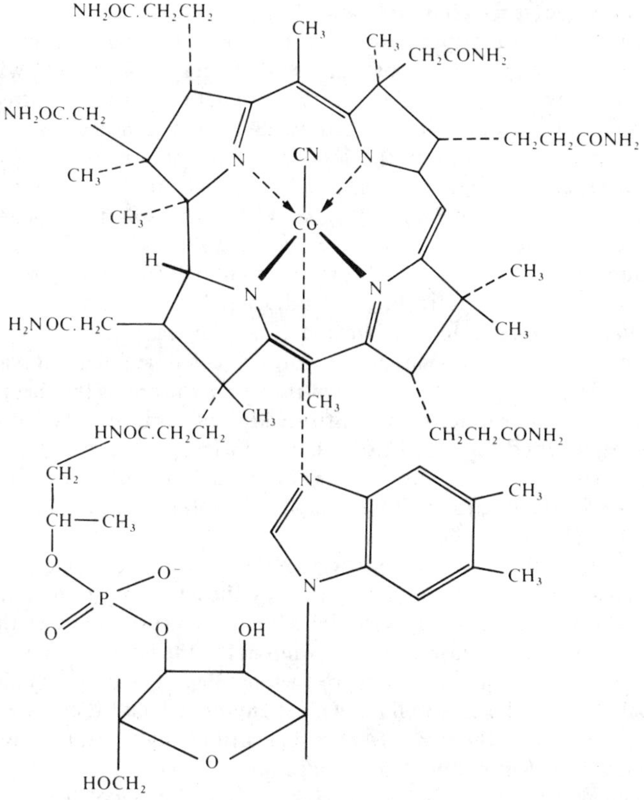

FIG.24. Vitamin B_{12} (cyanocobalamin). The cyanide group is an artefact of the isolation procedure. Note the similarity of the corrin ring surrounding the cobalt atom to the porphyrin structure in chlorophyll (Fig.14) and in haem (Fig.20).

† The process of formation of the erythrocytes.
‡ An increase in the number of erythrocytes.

the active constituent was isolated and was then shown to be a cobalt complex. It was given the name Vitamin B_{12}; the structure is shown in Fig.24.

The structure of the coenzyme has now been elucidated and Vitamin B_{12} consists of a highly substituted porphyrin-like corrin ring in which a Co^{3+} ion is bound to the four nitrogen atoms. The metal ion is additionally bound to the nitrogen of a nucleotide base which is attached to the sugar moiety by an unusual β-glycosidic linkage. In addition to the normal 5,6-dimethylbenzimidazole shown in Fig. 24, other bases such as adenine, 2-methyladenine, and guanine have been found in naturally occurring Vitamin B_{12} analogues.

This part of the structure which is reminiscent of many haem-iron compounds, is known as cobalamin, so that Vitamin B_{12} itself, which contains a cyanide group in the sixth coordination position is often known as cyanocobalamin. This compound is now synthesized commercially using various species of *Propioni* bacteria, replacing earlier methods using *Streptomyces.* The cyanide ion has no functional significance but is merely an artefact of the isolation procedure and can be displaced by many other ligands.

Subsequent studies showed that injections of Vitamin B_{12} were extremely effective in treating Coast disease. The effect on the erythrocyte count could be detected in as little as six hours. Following this it was quickly shown that the inorganic cobalt supplement was converted to Vitamin B_{12} by the bacterial microflora of the sheep's rumen and it was only when constituted in this form that the cobalt had the required activity. This explains why inorganic cobalt produces mainly toxic effects in both suckling lambs and monogastric animals, neither of which have developed suitable bacterial colonies in their digestive systems.

Inorganic cobalt is also beneficial in the treatment of some refractory anaemias, but unlike Vitamin B_{12} therapy the administration of cobalt salts always produces polycythaemia. It has been suggested that the cobalt may act by causing hypoxia (anoxia) in the bone marrow and this may well be the case, as the oxygen-binding capacity of simple cobalt(II) complexes is well known. At any rate, cobalt therapy produces exactly the same effect as high-altitude hypoxia, namely to stimulate the formation of erythrocytes.

The average human being contains a total of 2-5 mg of Vitamin B_{12} and its derivatives, concentrated in the liver. This is equivalent to ~ 1 p.p.m. Vitamin B_{12} and explains why it was so difficult to isolate originally. Vitamin B_{12} has been implicated in protein synthesis, the storage of folic acid, the activation of thiol enzymes, and myelin formation. However, its principal role is in the maturation of the erythrocyte. In the absence of Vitamin B_{12} the pro-erythroblast fails to reach the normoblast stage and instead is released as the immature megaloblast. These cells lose the nucleus and are found in the peripheral circulation as the macrocytes. Macrocytic anaemia (otherwise called pernicious or Addisonian anaemia) responds rapidly to treatment with Vitamin B_{12} supplements and the erythrocyte count is rapidly restored to normal without producing polycythaemia.

The normal human diet contains between 5 μg and 15 μg of Vitamin B_{12} daily, of which about 5 μg is absorbed. This has been demonstrated using radioactive cobalamin labelled with ^{58}Co(a β^+ emitter; half-life 71 d). This is probably the most useful cobalt isotope for tracer purposes since it is a weak β emitter and is available at high specific activities of up to 300 μCi per μg. ^{57}Co (E.C.; half-life 72 d) is less commonly used, and ^{60}Co (a β^- emitter; half-life 5·26 years) is normally employed as a γ radiation source in the radiotherapy of tumours, rather than as a tracer, since it is hepatotoxic.

The actual absorption of the cobalamin is a complex process involving a specific protein carrier. This protein (the intrinsic factor) must combine with Vitamin B_{12} (the extrinsic factor) before the vitamin can be absorbed, and this complex seems to act as a carrier, as it is found in an intact form in the tissues. The turnover of the cobalt complex is slow and appears to depend on the exact form of the cobalamin. Hydroxocobalamin has the longest half-life, about one year, whereas cyanocobalamin is cleared almost twice as rapidly.

Aquo-cobalamin, and its analogues with other nucleotides, has a water molecule in the sixth coordination position. These compounds are the active coenzymes for at least three biosynthetic reactions: the synthesis of methionine and acetate and methane formation. In each case the aquo-corrinoid compound appears to undergo transient methylation at the cobalt atom and so far the Vitamin B_{12} coenzymes are the only naturally occurring compounds known to contain a metal-carbon bond.

The second class of Vitamin B_{12} coenzymes also contain a metal-carbon bond. In these the cyanide ion is replaced by the 5′-deoxyadenosyl group which is bound to the metal by the carbon atom of the 5′-position (Fig.24). It now seems probable that the adenosyl group is introduced by reduction of the cobalamin to Co^{I}, using ferredoxin as the reducing agent followed by reaction with ATP in the presence of a suitable enzyme.

In vitro neutral or alkaline solution of cyanocobalamin can certainly be reduced to give both Co^{II} and Co^{I}-containing species called Vitamin B_{12r} and Vitamin B_{12s} respectively. The latter compound readily loses the CN^- from the sixth coordination position and thus produces a five-coordinate intermediate. Vitamin B_{12s} is a powerful reducing agent and will react even with water to give Vitamin B_{12r}, giving hydrogen in the process. It will also react with alkyl halides or acetylenes to give a number of organometallic complexes, so the suggested mode of incorporation of the adenosyl fragment is mechanistically quite feasible.

TABLE 9

Some reactions involving cobalt 5-deoxyadenosylcorrinoid coenzymes

R_1	R_2	R_3	Enzyme
H	$CH(NH_2)COOH$	COOH	Glutamate mutase
H	CO-CoA	COOH	Methylmalonyl-CoA mutase
CH_3	OH	OH	Dioldehydrase
CH_2OH	OH	OH	Glycerol dehydrase
H	NH_2	OH	Ethanolamine deaminase
H	NH_2	$CH_2CH(NH_2)CH_2COOH$	β-Lysine isomerase.

A good many enzymic reactions require this type of cofactor but all except one can be represented as a simple 1,2 hydride shift coupled with the reverse movement of a second group. Table 9 lists some of these.

$$R_1-\overset{R_2}{\underset{H}{C}}-\overset{H}{\underset{H}{C}}-R_3 \rightleftarrows R_1-\overset{H}{\underset{H}{C}}-\overset{R_2}{\underset{H}{C}}-R_3$$

The exception is the reaction catalysed by ribonuclease reductase but even this can be discussed in the same manner with the distinction that the hydrogen donor and acceptor are contained in separate molecules.

Typical of the 1,2 intramolecular shift is the diol-dehydrase reaction, which has been particularly well studied. If tritiated propylenediol CH_3.CHOH.CTHOH is employed, the coenzyme which is subsequently isolated contains tritium specifically substituted in the 5′-position of the adenosyl group. Clearly, therefore, the coenzyme acts as a hydrogen carrier and binding to the cobalt must activate the 5′-position for just this purpose.

One point remains unresolved. The coenzyme can be written as cobalt(III)-R^- in which a carbanion is bound to Co^{3+} ion or alternatively as Co^I-R^+, in which carbonium ion is bound to univalent cobalt and at the present time it is still uncertain which formation is 'correct'. It may even be that both are applicable in different situations; the problem will certainly repay further study.

Considering the complexity of Vitamin B_{12} it is surprising that its behaviour can be reproduced by a number of very simple model systems. The best known of these is cobalt(II) dimethylglyoximate which produces 'cobaloximes'. These are reduced to give unusual blue or green cobalt(I) compounds which, in the presence of tertiary phosphines, give hydrido-compounds $[H\text{-}Co^I(DMGH)_2(R_3P)]$. Such species react with alkyl halides or alkenes to give a variety of σ-bonded metal-alkyls.

Apart from the important role of the Vitamin B_{12} coenzyme, cobalt appears to have little biological function. Normally the level of free cobalt in the blood is less than 20 p.p.b. and this makes it very difficult to induce cobalt deficiency or to determine precisely what the metabolic role is, if any. It is known that *in vitro* at least, many metal-requiring enzymes are activated by Co^{2+} but since they also respond to other divalent metal ions such as Mg^{2+} and Zn^{2+} it is not certain whether they show any strict requirement for Co^{2+} *in vivo* – it seems unlikely. The likeliest possibility for Co^{2+} requirement is glycylglycinedipeptidase, which has been studied intensively *in vitro*, both directly and via model systems using cobalt(III) rather than cobalt(II) complexes (Fig.25).

Inorganic cobalt is mildly toxic, it depresses thyroid activity and a high cobalt diet may cause goitre. A more serious effect of cobalt toxicity is

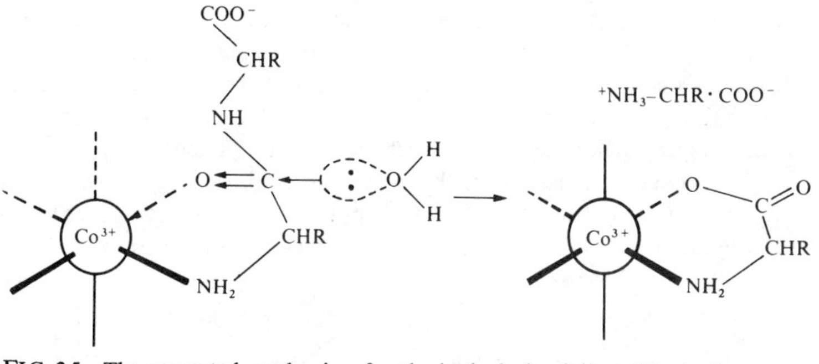

FIG.25. The suggested mechanism for the hydrolysis of dipeptides in the presence of cobalt(III). In the enzyme, cobalt(II) is thought to play a similar role. Note also the resemblance between this and the function of zinc in carboxypeptidase (Fig.16).

cardiomyopathy, progressive heart failure caused by excess glycogen. The cobalt, as Co^{2+}, binds lipoic acid thus preventing glycogen catabolism. This condition was observed in a number of heavy drinkers in Montreal and it was eventually traced to cobalt salts used as foam stabilizers in beer.

Nickel ($3d^8\ 4s^2$)

Only nickel(II) has any importance under conditions which are likely to be relevant biochemically, emphasizing the decreased stability of higher oxidation states towards the end of the first transition series. Nevertheless, the chemistry of nickel(II) is often bewilderingly complicated as its complexes may have a variety of coordination numbers and geometries. In many cases a single complex may exist in two or more forms which are linked in an equilibrium controlled by temperature, solvent, or even concentration. Equilibria of this type would seem to offer enormous possibilities in terms of fine control over metabolic activity, particularly in enzymic reactions where they could be utilized in allosteric control mechanisms and some evidence is now beginning to emerge for a functional role for nickel, though mainly connected with DNA and RNA metabolism.

Copper ($3d^{10}\ 4s^1$)

In older textbooks it is not uncommon to find copper treated along with Group 1A, to which a formal similarity in valence configuration suggests it might belong. Such comparisons are really rather strained since the alkali metals all form typically ionic compounds, whereas copper(I), by virtue of the filled-shell, d^{10}-configuration, is much more polarizable and hence is more likely to form covalent bonds (cf. Zn^{2+} and Ca^{2+}). Obviously compounds of copper in this oxidation state are diamagnetic and, since there can be no ligand-field transition, they are likely to be colourless, except where charge-transfer bands of suitable energy may arise. Numerous compounds containing copper(I) are known but they are usually stable only in the solid state or in non-aqueous solvents as the unipositive metal ion readily disproportionates in the presence of water. The driving force for this reaction comes primarily from the greater solvation energy of the copper(II) complexes. Calculations based on the appropriate oxidation potentials show that the maximum concentration of copper(I) is unlikely to exceed 10^{-2} M

$$Cu^+ + e \rightarrow Cu^0 \qquad E^\ominus = 0{\cdot}52\,V.$$

$$Cu^{2+} + e \rightarrow Cu^+ \qquad E^\ominus = 0{\cdot}15\,V.$$

$$2Cu^+ \rightarrow Cu^0 + Cu^{2+} \qquad E^\ominus = 0{\cdot}37\,V.$$

The free energy for this reaction $\Delta G = -nFE$ but the equilibrium constant

$$K_{eq} = -\log \frac{\Delta G}{RT},$$

where

$$K_{eq} = \frac{[Cu^{2+}]}{[Cu^{+}]^2}$$

so that $K_{eq} = 10^{-6}$. Hence, even if $[Cu^{2+}] = 10$ mole dm^{-3} then $[Cu^{+}] = 10^{-2.5}$ mole dm^{-3}.

Clearly such a disproportionation would be rather disastrous to the 'valence shuttle' mentioned later, but fortunately the stability of the unipositive oxidation state can be increased, either by introducing ligands which alter the oxidation potential or by changing the solvent. For example, ammonia increases the stability of copper(I) so that:

$$Cu^{2+} + Cu_{(metal)} \xrightarrow[NH_3(aq)]{} 2Cu^{+}.$$

Alternatively, solvents such as acetonitrile ($CH_3.CN$), which preferentially solvate the Cu^{+} ion may be used. Indeed in this solvent copper(II) is unstable and acts as a powerful one-electron oxidant.

The formation of the Cu^{2+} ion involves disturbing the d^{10}-configuration, which at first sight seems somewhat surprising, but as indicated previously the stability of the copper(II) state depends largely on the greater hydration energy of the +2 cation. By virtue of the higher charge and lower ionic radius this ion is less easily distorted than Cu^{+} and in consequence it forms a wide variety of stable complexes with sulphur, nitrogen, and oxygen donor ligands, but has little tendency to bind π-bonding ligands. The d^9 electronic configuration of Cu^{2+} can be arranged in only one way in the octahedral splitting pattern, namely $t_{2g}^6 e_g^3$, and this configuration leads to a Jahn-Teller distortion which has a considerable effect on the stereochemistry of copper(II) complexes. Hexaquo-copper(II), for example, has four short and two long Cu-O bonds. This tetragonal distortion makes the magnetic and spectroscopic properties of the copper(II) ion quite difficult to interpret. The ligand-field spectra should consist of only one band, as the d^9-configuration is formally equivalent to d^1 (the electron-hole equivalence) but instead it contains several overlapping bands arising from the multiple splitting of the d-orbitals.

The magnetic properties of copper(II) complexes are often unusual. Simple monomeric complexes usually have moments of 1·75 - 2·20 μ_B, slightly in excess of the spin-only value, but these complexes are in the minority. Deviations are often observed, particularly when measurements are made in the solid state, where metal-metal bonding may become important. The structure of copper(II) ethanoate (acetate) shows that it is actually a dimer in the solid state with the two copper atoms held in close proximity by bridging acetate groups so that the two unpaired e_g electrons can interact, pair their spins, and hence reduce the magnetic moment to the observed 'sub-normal' value. This type of interaction is likely to be significant in some of the enzymes discussed later which have several atoms of copper per enzyme molecule.

The limit of the Jahn-Teller distortion is of course the pseudo square-planar geometry which is favoured by many neutral complexes of copper(II). The spectroscopic and magnetic properties of such complexes are quite well understood. Anionic complexes such as $[Cu^{II}Cl_4]^{2-}$ usually adopt a (distorted) tetrahedral geometry but this is quite rare amongst the low molecular weight neutral complexes. It does, however, assume considerable importance in enzymic environments.

The terrestrial distribution of copper

The geochemistry of copper is well understood and its distribution and occurrence reflect its chalcophilic nature. The principal form of copper ores are the mixed iron-copper sulphides (chalcopyrites) and, like the iron, the copper is mobilized by the oxidation of sulphide to sulphate. The copper(II) salts thus formed are less easily precipated than the iron(II)/iron(III) species so that the soils generally have a lower copper content (average 25 p.p.m.) than the parent rock, due to leaching. As a result copper-deficient soils are widespread. Conversely, the copper-rich soil adjacent to natural deposits or mine-tailings provide areas of botanical interest, as the flora rapidly develops a specific copper-tolerance. There are, however, marked physiological effects on the animals grazing on such areas, though the pastures themselves are usually so localized as to be unobjectionable. The same cannot be said for the copper burden induced by man with the use of copper-based insecticides and fungicides. Repeated use of such pest-control agents can easily raise the available soil copper to dangerous levels.

A similar problem has been observed after using manure from intensively reared pigs which are fed with a copper supplement to increase the rate of weight-gain; the copper is poorly absorbed and the bulk of it passes through the animal to give unacceptably high levels in the manure.

The biological role of copper

Copper is widely distributed in both plants and animals and it is usually to be found linked to a variety of catalytic proteins. In many respects the biochemistry of copper resembles or even overlaps with that of iron since both are intimately involved in the metabolism of molecular oxygen. Thus the oxygen carrying pigment of mollusc blood is not haemoglobin but cuproprotein (haemocyanin). Copper also forms a vital part of a number of oxidase enzymes, sometimes in combination with other metal ions, as in cytochrome-c oxidase, which contains both copper and haem-iron.

The average adult human being contains between 100 mg and 150 mg copper, almost entirely protein-bound. This body pool is sustained by absorbing between 5 per cent and 10 per cent of the daily intake, which varies between 3 mg and 5 mg. Whilst the absorption of copper does not appear to be an active process it is certainly assisted by amino acids, and copper is actually transported across the membranes of the intestine as complexes of the type $[Cu^{II}(AA)_2]^{n+}$ (where AA = amino acid). Transport is most efficient with essential, acidic L-amino acids and least efficient with non-essential, basic D-amino acids. Once in the blood-stream the greater part of the copper is bound to the protein ceruloplasmin, molecular weight 151 000, in a complex containing eight Cu^{2+} ions per molecule. The remaining small portion of the copper is to be found in a ternary (Cu^{2+}-histidine-threonine) complex. Wilson's disease (the congenital absence of ceruloplasmin) results in copper accumulation which eventually proves fatal, though this condition has been treated with some success using chelating agents such as D-penicillamine to complex the unwanted copper ions so that they can be removed in the normal process of excretion.

Copper deficiency is well known to induce anaemia even if supplies of iron are more than sufficient, and it is thought that the copper is essential for the maturation of the erythrocytes and also for the incorporation of the iron into haem, as well as its mobilization from ferritin to transferrin. Copper deficiency in other animals is also thoroughly documented and is of considerable commercial importance to sheep farmers in Southern Australia where it is endemic. In this region the copper content of the soil is particularly low, leading to a severe form of ataxia known as 'swayback', from the characteristic drunken gait of the afflicted lambs. A less severe deficiency produces wool which lacks the characteristic kink or crimp and is of low market value, but fortunately both these conditions can be cured by supplementation with copper salts.

A wide variety of cuproprotein enzymes, together with a number of other copper-binding proteins whose functions are not yet clear, have now been isolated and purified. Generally these copper-protein complexes function as oxidases, using molecular oxygen as an electron acceptor. However, not all oxidases contain copper; many use flavin as the prosthetic group. It is currently believed that these cuproprotein enzymes operate by means of a redox cycle, the so-called 'valence shuttle' hypothesis, which involves oxidation of the substrate by copper(II) and subsequent regeneration of this ion from the copper(I), using molecular oxygen as the oxidizing agent (electron acceptor). This mechanism is not confined to enzymic reactions and the copper(II)/copper(I) cycle is also responsible for the copper-ion catalysis of the oxidation of ascorbic acid (Vitamin C). It is interesting to note that when this reaction also occurs *in vivo* where it is catalysed by ascorbate oxidase, the enzyme is about one thousand times more efficient than simple copper ions. Therefore, it would seem that in order to understand the enzymic function of copper it is necessary to grasp the factors which might affect the stability of these two oxidation states. (Copper is also known in the +3 oxidation state, where it is an even stronger oxidizing agent than copper(II). Copper(III) is unstable in aqueous solution but its intervention in enzymic oxidations should not be dismissed out of hand.)

Tyrosinase, the enzyme responsible for the oxidation of tyrosine and the production of the pigment melanin, has been isolated from *Neurospora crassa* and shown, by analysis, to contain one atom of copper per molecule of enzyme. The copper cannot be detected by e.s.r. spectroscopy and is formulated as copper(I). In support of this view it has been shown that tyrosinase is inhibited by CO or by CN^- ions, both of which are well known to bind to copper when it is in this oxidation state. The admission of oxygen in the absence of substrate causes a change to the divalent state but if substrate is added no copper(II) can be detected.

Many cuproproteins are amine oxidases catalysing the reaction:

$$R{-}CH_2{-}NH_2 + H_2O + O_2 \rightarrow R{-}CHO + H_2O_2 + NH_3.$$

Several of these copper-protein complexes are a beautiful bright blue colour when isolated in a pure state. The extinction coefficient of the copper chromophore in these complexes is at least ten times larger than that of simple copper-peptide complexes, and it is this which accounts for the intensity of the colour. At one time the unusual nature of the chromophore was thought to be the result of a copper-copper interaction, but the isolation from *pseudomonas fluorescens* of the blue curproprotein, *azurin*, which contains only one atom of copper per molecule has ruled out this suggestion. Although the possibility of a charge-transfer contribution to the colour must also be considered, it is now believed that the colour arises from the binding of the Cu^{2+} ions in a tetrahedrally distorted site which alters the splitting of the d-orbitals and, by a reduction in symmetry, increases the probability of a d-d transition occurring.

The distorted nature of the binding site is echoed both by the anomalous magnetic properties of the metal ion and also by the increase in redox potential of the Cu^{2+} from the value of 0·13 V, found in simple salts, to 0·33 V. A similar explanation was proposed for the copper(II)-biquinolyl complex which has an oxidation potential of +0·77 V. Clearly, a tetrahedral type of geometry will favour copper(I) which is known to prefer this type of coordination geometry. The polarity of the binding site may also have a considerable effect and, at this juncture, it is worthwhile recalling that some sites at the interior of the amino-acid residues can produce an environment almost as apolar as a liquid hydrocarbon. Consequently a copper ion incorporated into a protein may be in a totally different environment to a copper ion in aqueous solution, with a consequent effect on the copper(II)/ copper (I) equilibrium.

The redox potential of the so-called 'blue' copper has an even higher value in other cuproproteins. In the enzyme laccase, isolated from the tree *Polyporous versicolor*, it reaches a value of +0·767 V and is the highest yet recorded for a transition-metal ion bound to a protein. Because of this wide range of redox potentials it may be unwise to generalize about the entire group of 'blue' cuproproteins. Analysis of *polyporous* laccase reveals that it contains a total of four atoms of copper per enzyme, and anaerobic titrations with reducing agents reveal that they must all be considered as copper(II). One of these is readily distinguished as a 'blue' copper(II) and another as a 'non-blue' copper(II). The 'non-blue' copper resembles the simple Cu^{2+} ion more closely than does the 'blue' form but there are still some important differences, not least is the avidity with which the 'non-blue' copper binds anions.

What is surprising is that when the enzyme is examined by e.s.r. spectroscopy only these two Cu^{2+} ions are detectable. It seems probable that the 'e.s.r. non-detectable' copper is genuinely copper(II), as redox titrations imply, and the two ions exist as a pair with a metal-metal interaction that is strong enough to result in spin-pairing and apparent diamagnetism. Of course this situation is just a further development of the copper interaction described previously for copper ethanoate. Such a situation is not confined to laccase and both ascorbic acid oxidase and ceruloplasmin contain eight atoms of copper per molecule, all of which are formally copper(II). These are divided into two 'blue', two 'non-blue' and four 'e.s.r. non-detectable'.

It appears that the combination of these three types of copper provides a particularly efficient way of performing oxidations and it seems likely that the high redox potential of the 'blue' copper(II) species suits it to be the electron acceptor from substrate. Unfortunately, however, little is known about the interaction of a substrate with these enzymes, though preliminary experiments indicate that the substrate binds to the surface of the protein and the electrons lost in oxidation are carried by the protein chain to the 'blue' copper. The reoxidation of the reduced 'blue' copper is carried out by the 'e.s.r. non-detectable' copper pair and it seems reasonable to suggest that these function as a pair in order to provide a two-electron acceptor for the oxygen molecule and thus overcome the thermodynamic disadvantage of the one-electron oxidation.

The remaining class of cuproprotein amine oxidases also require pyridoxal phosphate for activity. Typical of these is the enzyme isolated from pea seedlings. This is thought to contain copper as copper(II) since treatment with chelating agents removes both copper and enzymic activity, which may be fully restored by the addition of further dipositive copper. Studies with model systems containing copper(II), pyridoxal (or the carbocyclic analogue salicylaldehyde), and a variety of amines suggest that these enzymes function by the formation of a ternary complex. This contains the pyridoxal and amine as a Schiff's base and by varying the reaction conditions, these model complexes can be made to undergo a wide series of reactions which mimic biological transformations.

Molybdenum ($4d^5\ 5s^1$)

Inspection of Fig. 1 (p. 2) shows that molybdenum is the congener of chromium in Group VI and therefore it might be expected that the chemistries of the two elements should show some similarities. Certainly both are known in all oxidation states from -2 to +6 (as is to be expected, the lower valency states are found in complexes of π-acceptor ligands whereas the higher valency states are associated with oxide and fluoride complexes), but the relative stabilities of these states differ so widely between the two metals that for the most part it is probably best to ignore the supposed group similarity.

The relevant oxidation potentials emphasize this point most clearly. Thus, whereas chromium in an aqueous environment forms chromium(III) complexes most readily, molybdenum is found either in the +5 or +6 states, and the formation of molybdenum(III) is energetically less favourable:

$$\mathrm{Cr^{VI}} \xrightarrow[\mathrm{H^+}]{+1{\cdot}33\ \mathrm{V}} \mathrm{Cr^{III}} \xrightarrow[\mathrm{H^+}]{-0{\cdot}78\ \mathrm{V}} \mathrm{Cr^{II}} \xrightarrow{-0{\cdot}33\mathrm{V}} \mathrm{Cr(metal)}$$

$$\mathrm{Mo^{VI}} \xrightarrow[\mathrm{H^+}]{-0{\cdot}6\ \text{to}\ -1{\cdot}0\ \mathrm{V}} \mathrm{Mo^{III}} \xrightarrow[\mathrm{H^+}]{-0{\cdot}2\ \mathrm{V}} \mathrm{Mo(metal)}.$$

The simple low molecular weight complexes of molybdenum(V) and (VI) are usually anionic and often incorporate oxo- (O^{2-}) ligands. The molybdenum(V) species have a strong tendency to dimerize, forming diamagnetic species by the pairing of the singled-electron on each metal ion. Otherwise, molybdenum(V) species are paramagnetic, with simple magnetic properties reflecting the one unpaired electron. Molybdenum(VI) complexes, on the other hand, are diamagnetic, as the formal $4d^0$ configuration would predict. They too have a tendency to dimerize.

The exception to these remarks seems generally to be where molybdenum is bound to thiolate (RS^-) ligands, when molybdenum(V) is often reduced with the concomitant oxidation of the ligand:

$$\mathrm{Mo^{V}} + 2\mathrm{RS^-} \rightarrow \mathrm{R{-}S{-}S{-}R} + \mathrm{Mo^{III}}.$$

Since the redox potential of such ligands is only about -0·30 V, the potential of the Mo^{V}/Mo^{III} couple must be considerably reduced by complex-formation with sulphur donor ligands so that the possibility that molybdenum(III) might occur in enzymic environments cannot be ignored.

The terrestrial distribution of molybdenum

Molybdenum is one of the rarer elements in the earth's crust, being 54th in the list of abundance. Typical levels in the soil are about 2 p.p.m., but around molybdenum mineralizations this may be higher. Local deposits of molybdenite (molybdenum disulphide) are often indicated by yellow-orange soil colouration and stunting of plant growth. Since molybdenum is readily precipitated as hydrated oxides or as insoluble salts of the hetero-polyanions† the concentration in ground-water is usually very low.

The biological role of molybdenum

The molybdenum requirement of most organisms is very low. Thus nitrogen-fixing bacteria grow well in solution containing less than 0·1 p.p.m. of molybdenum. Nevertheless deficiency symptoms are sometimes encountered. Molybdenum deficiency in pasture land is not uncommon and the application of small amounts of simple molybdenum salts has beneficial effects on the yield of herbage. Some plants actually accumulate this element, the alder being particularly well known in this respect. The ecological implications are surprising: in one case it was recorded that alders growing on the shores of a California lake reduced the concentration of molybdenum in the waters sufficiently to interfere with the growth of plankton and hence reduce the numbers of fish.

Molybdenum-copper interactions are particularly important and excessive levels of one element in the nutrients of either plants or animals inhibit the uptake of the other. Thus herbivores grazing on local-high-molybdenum pastures exhibit characteristic symptoms caused by copper deficiency rather than an excess of molybdenum, whilst excess dietary copper results in disturbances of purine metabolism due to molybdenum deficiency interfering with the level of active enzymes of the xanthine-oxidase type.

Molybdenum is required by most organisms, including man, as a cofactor for a variety of enzymes. In each case the metal plays the same role, acting as either a donor or acceptor of electrons in redox reactions. The electron-transfer process between the molybdenum and the substrate may occur directly but more often it involves the iron atoms which are found so often in molybdenum-requiring enzymes. These enzymes can be neatly divided into two classes, the reductases and the oxidases, but in both cases a similar

† Complex anionic oxides containing more than one molybdenum atom.

mechanism is believed to operate, in which the molybdenum cycles between the +5 and +6 oxidation states.

The reductases are all involved either in the process of nitrogen fixation or the reduction of nitrate, and hence are of the utmost importance since plants, and consequently animals, obtain their nitrogen by one of these two routes. The direct fixation of nitrogen is carried out by many organisms, but the most closely studied have been *Clostridium pasteurianum* and *Azotobacta vinelandii*. From these and other organisms have been isolated the molybdenum-containing proteins responsible for nitrogen fixation (nitrogenase); depending on the condition of purification the molybdenum is in either the +5 or +6 oxidation state. This can be shown by treatment of the nitrogenase enzyme with chelating agents such as 8-hydroxyquinoline which readily forms complexes with molybdenum in either of these states. More directly, the oxidation state, and the change in oxidation state of the molybdenum during the reduction cycle, can be studied by observation of the electron spin resonance spectrum of the molybdenum(V). The nature of the signals observed during the cycle of enzyme action are unlike those previously observed in low molecular weight complexes and they show quite clearly that the molybdenum ions are in an 'unusual' environment; this no doubt accounts for their rather unique properties.

The molybdenum ions are believed to act by a 'valence shuttle' mechanism involving reversible transfer between +5 and +6 oxidation states with the subsequent transfer of the electron(s) to the nitrogen molecule. However, the reaction sequence has not yet been completely clarified. Indeed, it is increasingly apparent that it is far from simple and it is believed to involve several iron enzymes and coenzymes, as shown in Fig.26.

Nor is the detailed mechanism of nitrogenase action yet settled. It still remains to be seen if the two molybdenum atoms common to most of the enzymes act as a pair as has been suggested, or if they function independently. Certainly, their joint operation seems an attractive hypothesis to explain the ease with which nitrogen is reduced enzymically. This mode of action has also been suggested to operate in certain other molybdenum-containing enzymes, the reductases.

As an alternative to the direct reduction of the nitrogen molecule, the incorporation of nitrogen into plants may occur via the reduction of nitrate (hence the use of nitrate fertilizers). The first stage of this assimilatory nitrate reduction in micro-organisms and higher plants is the reduction of nitrate to nitrite. It has been shown that the molybdenum ions provide a link between nitrate and FAD in this scheme, and that two molybdenum ions are necessary to perform the two-electron transfer (Fig.26).

A second class of nitrate reductases exists, with a different function. These respiratory or dissimilatory nitrate reductases utilize nitrate as a terminal electron acceptor in place of oxygen without utilizing the products. It is thought that the overall mechanism of the reaction is

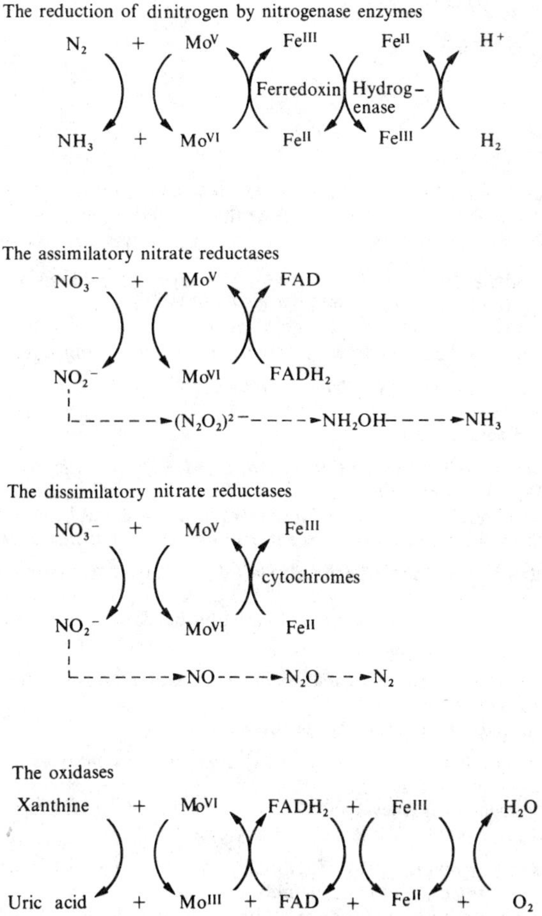

The reduction of dinitrogen by nitrogenase enzymes

N_2 + Mo^{V} Fe^{III} Fe^{II} H^+

Ferredoxin Hydrogenase

NH_3 + Mo^{VI} Fe^{II} Fe^{III} H_2

The assimilatory nitrate reductases

NO_3^- + Mo^{V} FAD

NO_2^- Mo^{VI} $FADH_2$

$\rightarrow (N_2O_2)^{2-} \rightarrow NH_2OH \rightarrow NH_3$

The dissimilatory nitrate reductases

NO_3^- + Mo^{V} Fe^{III}

cytochromes

NO_2^- Mo^{VI} Fe^{II}

$\rightarrow NO \rightarrow N_2O \rightarrow N_2$

The oxidases

Xanthine + Mo^{VI} $FADH_2$ + Fe^{III} H_2O

Uric acid + Mo^{III} + FAD + Fe^{II} + O_2

FIG.26. The 'valency shuttle' in the action of molybdenum reductases and oxidases.

similar to that of the assimilatory enzymes with the extra involvement of one or more cytochromes between the FAD and the molybdenum.

The molybdenum-requiring oxidases utilize either xanthine (and purines) or aldehydes as substrates. These are believed to operate by cycling between Mo^{VI} and Mo^{V}, that is to say, in the opposite direction to the reductases. A general reaction is shown in Fig.26. More recently, in order to explain the detailed mechanism of the reaction, it has been suggested that the Mo^{VI} may in fact be reduced to Mo^{III}; clearly the presence of cysteine residues at the active site will be favourable to such a reaction. The evidence currently available suggests that this might indeed be the case.

Further Reading

THE range of textbooks dealing with inorganic chemistry and with biochemistry is huge, and almost all will provide much of the basic information required. Those included here are simply ones with which the author is personally familiar.

Best, C. H. and Taylor, N. B. (Eds.) (1966). *The physiological basis of medical practice* (8th edn). Williams and Wilkins, London.

An enormous book providing an immense amount of information, but highly readable. Careful use of the index can make this a useful reference work.

Bowen, M. J. M. (1966). *Trace elements in biochemistry.* Academic Press, New York.

A most useful compendium of data.

Cotton, F. A. and Wilkinson, G. (1972). *Advanced inorganic chemistry* (3rd edn). John Wiley and Sons, London.

This provides a much higher level of treatment, but it will still be useful to the non-specialist as a reference work since it contains a wealth of material.

Goldschmidt, V. M. (1958). *Geochemistry* (Ed. A. Muir). Clarendon Press, Oxford.

Though no longer young this is still undoubtedly the classic refence book on this subject.

Hughes, M. N. (1972). *The inorganic chemistry of biological processes.* John Wiley and Sons, New York.

A more advanced treatment of the role of metal ions.

Lowrie, R. S. and Ferguson, H. J. C. (1971). *Inorganic and physical chemistry.* Pergamon Press, Oxford.

This book, designed for A-level and HNC courses, should provide an explanation of chemical principles for the less chemically orientated reader.

Mazur, A. and Harrow, B. (1971) *Textbook of biochemistry.* (10th edn). W. B. Saunders Company, Philadelphia.

A thorough account of basic biochemical principles with a useful bias towards physiological cause and effect.

Stalfelts, M. G. (1972). *Stalfelts plant ecology.* (translated by M. S. Jarvis and P. G. Jarvis). Longmans, London.

An elegant account of the many interactions betwen plant and environment, but particularly those between plant and soil.

Sutcliffe, J. F. and Barker, D. A. (1974). *Plants and mineral salts.* Edward Arnold, London.

A brief but highly informative account of this aspect of plant physiology.

Underwood, E. J. (1971). *Trace elements in human and animal nutrition* (3rd edn). Academic Press, New York.

A detailed account of the phenomenology of the nutritional and toxicological effects of trace elements, but with little chemical bias.

Williams, D. R. (1971). *The metals of life.* Van Nostrand, New York.
An account of the roles of metals concentrating particularly on the effects and energetics of complex-formation.

Related books in the Oxford Chemistry Series

R. J. Puddephatt: *The periodic table of the elements* (OCS 3).

A. Earnshaw and T. J. Harrington: *The chemistry of the transition elements* (OCS 13)

G. Pass: *Ions in solution (3): inorganic properties* (OCS 7).

W. S. Fyfe: *Geochemistry* (OCS 16).

D. J. Spedding: *Air pollution* (OCS 20).

C. F. Bell: *Principles and applications of metal chelation* (OCS 25)

Index